PRESERVED LOCOMOTIVES
OF BRITISH RAILWAYS

TWELFTH EDITION

Peter Fox, Peter Hall & Robert Pritchard

PLATFORM 5

Published by Platform 5 Publishing Ltd., 3 Wyvern House, Sark Road, Sheffield. S2 4HG, England.
Printed in England by Wyndeham-Gait, Grimsby

ISBN 9 781902 336572

© 2007 Platform 5 Publishing Ltd. All rights reserved. No part of this publication may be reproduced or transmitted in any form or by any means electronic, mechanical, photocopying, recording or otherwise, without the prior permission of the publisher.

Above: LMS Class 7F 0-8-0 (ex-LNWR Class G2) No. 49395 pulls a load of box vans at Quorn on the Great Central Railway on 15 January 2007. **John Calton**

Front Cover Photograph: 3700 "CITY" Class 3440 CITY OF TRURO is seen near Castle Hill on the West Somerset Railway with a Minehead–Bishops Lydeard service on 18 March 2005.
Rail Photoprints

Back Cover Photograph (top): BR Green-liveried 20214 leaves Newby Bridge on the Lakeside & Haverthwaite Railway with an afternoon service for Haverthwaite on 13 November 2005.
Tom Heavyside

Back Cover Photograph (bottom): A mixed DMU set formed of Class 104 DMBS's 50528+50454 and Class 127 DMBS 51618 stands at Carrog during the Llangollen Railway's DMU gala on 10 June 2006.
Paul Chancellor

CONTENTS

Forward to the Twelfth Edition 4
Introduction 5
General Notes 5
1. Steam Locomotives 6
    1.1. Great Western Railway & Absorbed Companies' Locomotives 7
    1.2. Southern Railway & Constituent Company Locomotives 21
    1.3. London Midland & Scottish Railway & Constituent Company Locomotives 33
    1.4. London & North Eastern Railway & Constituent Company Locomotives ... 43
    1.5. British Railways Standard Steam Locomotives 68
    1.6. War Department Locomotives 72
    1.7. United States Army Transportation Corps Steam Locomotives 75
    1.8. Replica Steam Locomotives 76
    1.9. Overseas Steam Locomotives Preserved in Great Britain 78
2. Diesel Locomotives 84
    2.1. London Midland & Scottish Railway Diesel Locomotives 85
    2.2. Southern Railway Diesel Locomotives 86
    2.3. British Railways Diesel Locomotives 87
    2.4. Experimental Diesel Locomotives 105
    2.5. Departmental Diesel Locomotives 105
3. Electric Locomotives 107
    3.1. Pre-Nationalisation Design Electric Locomotives 107
    3.2. British Railways Electric Locomotives 108
4. Gas Turbine Vehicles 111
5. Diesel Multiple Unit Vehicles 129
    5.1. GWR Diesel Railcars 130
    5.2. British Railways DMUs 130
6. Electric Multiple Unit Vehicles 143
    6.1. Southern Railway EMU Cars 143
    6.2. Pullman Car Company EMU Stock 145
    6.3. LMS & Constituents EMU Stock 146
    6.4. LNER & Constituents EMU Stock 147
    6.5. BR EMU Stock 148
7. List of Locations 154
    7.1. Preservation Sites & Operating Railways 154
    7.2. Public Houses & Hotels 157
8. Abbreviations Used 157
9. Private Manufacturers Codes 158

FOREWORD TO THE TWELFTH EDITION

At last, the new edition of this book has been published and the opportunity has been taken to make a number of changes. Firstly we have cut out a number of "peripheral" sections, such as military locomotives other than those of types used by main line companies, trolleys, locomotives confined to depots, workshops etc. to concentrate on the main purpose of the book which is to detail preserved main line locomotives and multiple unit vehicles of BR and constituents, along with overseas main line steam locomotives in Great Britain.

We have also decided not to include former main line locomotives in industrial service in Great Britain in this book. Such locomotives are now in a separate section in the Platform 5 "Locomotives" Pocket Book and in the combined edition "Locomotives and Coaching Stock 2008", published March 2008.

The opportunity has been taken to improve the technical information for steam locomotives particularly regarding valve gear and superheat with BR power classifications now being included for ex-GWR, SR, LNER and WD locomotives as well as for ex LMS and BR standard locos.

The building of new steam locomotives to existing designs is something which would have seemed to be science fiction a few years ago, but the A1 Locomotive Trust's new loco 60163 TORNADO is now nearing completion. In addition the West Somerset Railway have created a 4300 Class 2-6-0 out of a 5101 Class 2-6-2T. A number of new build projects have now been mooted and we have included details of these in separate sections at the end of each particular steam section. Some are completely new build, whilst others utilise parts of donor locomotives. How many of these locos will actually be built is a matter for conjecture, but we wish the organisations building these locos well.

Recent years have seen continual change in the railway preservation movement with many locomotives and multiple units on the move and further additions to the ranks of preserved diesel and electric locomotives and multiple unit stock, especially EMU vehicles.

On the diesel front, several main line locomotives once considered to be preserved have returned to commercial operation, whilst others of similar types which would once have passed to preservationists find themselves with smaller commercial operators. Such locomotives are not included, but will also be found in the Platform 5 "Locomotives" Pocket Book and in the combined edition "Locomotives and Coaching Stock 2008".

Preserved steam and diesel locomotives plus the odd DC electric have seen regular use on the main line network in recent years, but particular mention should be made of the AC Locomotive Group's 86101. This is the first AC electric to be used on the main line.

Thanks are given to all those who have knowingly and unknowingly assisted in the updating of this book. As the authors are not infallible, they would welcome notification of any corrections or updates to the book of which readers have first hand knowledge.

Please send comments or amendments to Robert Pritchard at the Platform 5 address on the title page or by e-mail to Robert@platform5.com (phone 0114 2552625).

Peter Fox. October 2007.

INTRODUCTION

This book contains details of preserved locomotives and multiple unit stock which has been in the ownership of, or operated under the jurisdiction of the British Railways Board, its constituents or its descendants. Only vehicles no longer in the ownership of the descendants of the British Railways Board or commercial railway companies are included.

In addition, overseas locomotives preserved in Great Britain are included, but with a minimum gauge of 750 mm, and a requirement that they must either have been locos of a national network, or preserved on an ex-BR line.

Also included are War Department steam locomotives and those of the United States Army Transportation Corps which are currently resident in Great Britain.

As an exception, ex-BR or WD locos in service abroad are included, plus 0-6-0 shunters built for industry to, or similar to, the WD design which became LNER J94.

The book is updated to information received by 3 October 2007.

GENERAL NOTES

This book has been divided into several main categories namely steam locomotives, diesel locomotives, electric locomotives, gas turbine vehicles and multiple unit stock. Further details applicable to each category being given in the introductory paragraph of each section.

Details regarding technical details of the various locomotives and multiple units can be found in the introductory paragraphs for each section. A few details are, however, consistent for each section, these being:

Numbers

All numbers carried at various times by the locomotives are shown, except in the case of temporary identity changes for filming and similar events. For each railway, numbers are given in chronological order. Details are also given of the current identity if it had not previously been carried.

Names and Class Names

Names bestowed after preservation are shown in inverted commas. Official names and class names are shown without inverted commas.

Locations

The location where the locomotive or multiple unit is normally to be found is given if known. Fuller details of locations in Great Britain including OS grid references are given under "List of Locations". It is not uncommon for locomotives and multiple units to visit other locations for operation, display or mechanical attention. Where such visits are of a long-term nature, the locomotive or multiple unit is shown as being at its host site with details of its home site being given as a footnote. Details of locomotives undergoing restoration away from their home site are also given as a footnote. (N) denotes a locomotive or multiple unit from the National Collection not currently located at the National Railway Museum York, or at the outpost at Shildon.

Build Details

For each locomotive the builder, works number (if any) and year of build are given. Private builder codes will be found in section 9.

Updates

Updates to this book are published in Platform 5's monthly magazine, **Today's Railways UK**, available from newsagents or on direct subscription (see inside covers).

1. STEAM LOCOMOTIVES

Almost from the birth of the railways to the 1960s, steam was the principal form of propulsion, its rapid decline in the 1960s being the impulse for much of the preservation movement. Steam locomotives are arranged generally in numerical order of the British Railways number, except that very old locomotives which did not receive numbers in the series pertaining at nationalisation in 1948 are listed at the end of each pre-nationalisation company section.

A select number of steam locomotives are permitted to work steam specials on the national railway network. Such locomotives invariably spend long periods away from their home bases undertaking such duties. Other locomotives may also spend periods away from their home bases as guests of other locations.

Wheel Arrangement

The Whyte notation is used for steam locomotives and diesel shunters with coupled driving wheels. The number of leading wheels are given, followed by the number of driving wheels and then the trailing wheels. Suffixes are used to denote tank locomotives as follows: T – side tank, PT – pannier tank, ST – saddle tank, WT – well tank.

Dimensions

These are given in imperial units for steam locomotives as follows:

Boiler pressure.	In pounds force per square inch (lbf/sq. in.).
Cylinders	Cylinder dimensions. The diameter is given first followed by the stroke. (I) indicates two inside cylinders, (O) two outside cylinders, (V) two vertical cylinders, (3) three cylinders (two outside and one inside) and (4) four cylinders (two outside and two inside).
Wheel diameters	These are given from front to back, i.e. leading, driving, trailing.
Weights	These are given in full working order.

Tractive Effort

This is is given at 85% boiler pressure to the nearest 10 lbf for steam locomotives. Phillipsons formula has been used to calculate these as follows:

$$T.E. = \frac{.85d^2snp}{2w}$$

where
T.E. = tractive effort in pounds force;
d = cylinder diameter (inches);
s = piston stroke (inches);
n = number of cylinders;
p = maximum boiler pressure (lb/sq. in.);
w = new driving wheel diameter (inches).

Brakes

Steam locomotives are assumed to have train vacuum brakes unless otherwise stated.

Valve Gear

Unless stated otherwise valve gear on steam locomotives is assumed to be inside the locomotive, except Walschaerts and Caprotti gears which are assumed to be outside. Exceptions are LMS-design Class 5MT 4-6-0 44767 which has outside Stephenson valve gear and LNER (ex-GER) Class N7 0-6-2T No. 69621 which has inside cylinders and Walschaerts valve gear and various 4-cylinder GWR classes which have inside Walschaerts valve gear with rocking shafts for the outside cylinders.

GWR

1.1. GREAT WESTERN RAILWAY AND ABSORBED COMPANIES' STEAM LOCOMOTIVES

GENERAL

The GWR was the only one of the big four companies which was essentially an existing company. A number of smaller companies were absorbed at the grouping. These were virtually all in Wales and included the Cambrian Railways, Cardiff Railway, Rhymney Railway and Taff Vale Railway. Prior to 1923 the GWR had also absorbed smaller concerns at various dates.

Numbering & Classification System

The locomotives of the absorbed companies were given the lower numbers and GWR classes the higher numbers. Instead of arranging classes in blocks, the GWR adopted a strange system whereby the second digit remained constant within a class, e.g. the 0-6-2Ts numbered 5600–99 continued with 6600–99. Sometimes earlier numbers were filled in, e.g. 5101–99 continued with 4100–99. Classes were always denoted by the number of the first member of the class to be built, which was not always the lowest number in the series. GWR locos were not renumbered by BR on nationalisation.

Power Class & Route Restriction System

The GWR adopted a power classification letter code system which ranged from A to E in ascending order of power. Certain small locomotives which were below group A were said to be unclassified and the "Kings" were classed as "special" being higher than "E". BR power classifications are also shown in brackets in this section.

Route restriction was denoted by a system of coloured spots painted on the cabside. In ascending order of restriction these were as follows: Yellow, Blue, Red, Double Red. Where no restriction is specified, locomotives were unrestricted.

2884 Class 2-8-0 No. 3802 runs round its stock at Llangollen on 10 June 2006.
Rail Photoprints

CORRIS RAILWAY   0-4-2ST
Built: 1878 by Hughes Locomotive & Tramway Engine Works Ltd. as 0-4-0ST. Rebuilt 1900 to 0-4-2ST.
Gauge: 2' 3".
Boiler Pressure: 160 lbf/sq. in.   **Weight:** 9 tons
Wheel Diameters: 2' 6", 10".   **Cylinders:** 7" x 12" (O).
Valve Gear: Stephenson.   **Tractive Effort:** 2670 lbf.

3   "SIR HAYDN"   Talyllyn Railway   HLT 323/1878 reb. 1901

CORRIS RAILWAY   0-4-2ST
Built: 1921.
Boiler Pressure: 160 lbf/sq. in.   **Gauge:** 2' 3".
Wheel Diameters: 2' 0", 1' 4½".   **Weight:** 8 tons.
Valve Gear: Hackworth.   **Cylinders:** 7" x 12" (O).
    Tractive Effort: 3330 lbf.

4   "EDWARD THOMAS"   Talyllyn Railway   KS 4047/1921

VALE OF RHEIDOL   2-6-2T
Built: 1923–24. GWR development of an older design built 1902.
Gauge: 1' 11½".
Boiler Pressure: 165 lbf/sq. in.   **Weight:** 25 tons.
Wheel Diameters: 2' 0", 2' 6", 2' 0".   **Cylinders:** 11½" x 17" (O).
Valve Gear: Walschaerts. Piston valves.   **Tractive Effort:** 10 510 lbf.

7   OWAIN GLYNDWR   Vale of Rheidol Railway   Swindon 1923
8   LLYWELYN   Vale of Rheidol Railway   Swindon 1923
9   PRINCE OF WALES   Vale of Rheidol Railway   Swindon 1923

No. 12   4wT
Built: 1926. Sentinel vertical-boilered geared locomotive. Returned to manufacturer after 3 months service.
Boiler Pressure: 275 lbf/sq. in.   **Weight:** 20 tons.
Wheel Diameter: 2' 6".   **Cylinders:** 6" x 9" (I).
Valve Gear: Rotary cam.   **Tractive Effort:** 7200 lbf.

GWR   Present
12   49 "No. 2 ISEBROOK"   Lavender Line, Isfield   S 6515/1926

OXFORD, WORCESTER & WOLVERHAMPTON RAILWAY   0-6-0
Built: 1855. Following withdrawal in 1904 the cut-down frames, motion and four-coupled wheels were set up as an instructional model. It has been preserved in this form although presently dismantled.
Boiler Pressure: 140 lbf/sq. in.   **Weight:** 32.60 tons
Wheel Diameter: 5' 3"   **Cylinders:** 17½" x 24" (I).
Valve Gear: Stephenson?   **Tractive Effort:** 13 100 lbf.

GWR   OW&WR
252   34   Leeds Industrial Museum   EW 1855

TAFF VALE RAILWAY CLASS O2   0-6-2T
Built: 1899 by Neilson, Reid. Sold by GWR 1926. 9 built.
Boiler Pressure: 160 lbf/sq. in.   **Weight:** 61.5 tons.
Wheel Diameters: 4' 6½", 3' 1".   **Cylinders:** 17½" x 26" (I).
Valve Gear: Stephenson. Slide valves.   **Tractive Effort:** 19 870 lbf.
Power Class: B.   **Restriction:** Blue.

GWR   TVR
426   85   Keighley & Worth Valley Railway   NR 5408/1899

TAFF VALE RAILWAY CLASS O1   0-6-2T

Built: 1894–97. Survivor sold by GWR 1927. 14 built.
Boiler Pressure: 150 lbf/sq. in.   **Weight:** 56.4 tons.
Wheel Diameters: 4' 6½", 3' 8¾".   **Cylinders:** 17½" x 26" (I).
Valve Gear: Stephenson. Slide valves.   **Tractive Effort:** 18 630 lbf.
Power Class: A.   **Restriction:** Yellow.

GWR	TVR		
450	28	Dean Forest Railway (N) Cardiff West Yard	306/1897

PORT TALBOT RAILWAY   0-6-0ST

Built: 1900/01. Survivor sold by GWR 1934. 6 built.
Boiler Pressure: 160 lbf/sq. in.   **Weight:** 44 tons.
Wheel Diameter: 4' 0½".   **Cylinders:** 16" x 24" (I).
Valve Gear: Stephenson. Slide valves.   **Tractive Effort:** 17230 lbf.
Power Class: A.   **Restriction:** Yellow.

GWR	PTR		
813	26	Severn Valley Railway	HC 555/1901

WELSHPOOL & LLANFAIR RAILWAY   0-6-0T

Built: 1903. 2 built.
Boiler Pressure: 150 lbf/sq. in.   **Gauge:** 2' 6".
Wheel Diameter: 2' 9".   **Weight:** 19.9 tons.
    Cylinders: 11½" x 16" (O).
Valve Gear: Walschaerts. Slide valves.   **Tractive Effort:** 8180 lbf.

BR	GWR	W&L			
822	822	1	THE EARL	Welshpool & Llanfair Railway	BP 3496/1903
823	823	2	THE COUNTESS§	Welshpool & Llanfair Railway	BP 3497/1903

§ Name altered to COUNTESS by GWR. Now renamed THE COUNTESS.

POWLESLAND & MASON   0-4-0ST

Built: 1903–06. Survivor sold by GWR 1928 for industrial use.
Boiler Pressure: 140 lbf/sq. in.   **Weight:** 24.85 tons.
Wheel Diameter: 3' 6".   **Cylinders:** 14" x 20" (O).
Valve Gear: Stephenson. Slide valves.   **Tractive Effort:** 11 110 lbf.

GWR	P&M		
921	6	Snibston Discovery Park	BE 314/1906

CARDIFF RAILWAY   0-4-0ST

Built: 1898. Rebuilt Tyndall Street Works 1916.
Boiler Pressure: 160 lbf/sq. in.   **Weight:** 25.5 tons.
Wheel Diameter: 3' 2½".   **Cylinders:** 14" x 21" (O).
Valve Gear: Hawthorn-Kitson.   **Tractive Effort:** 14 540 lbf.

GWR	Car.R.		
1338	5	Didcot Railway Centre	K 3799/1898

ALEXANDRA DOCKS & RAILWAY COMPANY   0-4-0ST

Built: 1897. Rebuilt Swindon, 1903. Sold by GWR 1932 for industrial use.
Boiler Pressure: 120 lbf/sq. in.   **Weight:** 22.5 tons.
Wheel Diameter: 3' 0".   **Cylinders:** 14" x 20" (O).
Valve Gear: Stephenson. Slide valves.   **Tractive Effort:** 11 110 lbf.

GWR		AD			
1340	TROJAN	TROJAN	Didcot Railway Centre		AE 1386/1897

▲ Welshpool & Llanfair Railway 2' 6" gauge 0-6-0T No. 823 THE COUNTESS at Welshpool on 20 March 2005.
Hugh Ballantyne

▼ GWR 1400 Class 0-4-2T No. 1450 propels an ex GWR auto coach towards Bishops Lydeard with a service from Norton Fitzwarren at the West Somerset Railway gala of 18 March 2005.
Rail Photoprints

1361 CLASS   0-6-0ST

Built: 1910. Churchward design for dock shunting. 5 built (1361–5).
Boiler Pressure: 150 lbf/sq. in.   **Weight:** 35.2 tons.
Wheel Diameter: 3' 8".   **Cylinders:** 16" x 20" (O).
Valve Gear: Allan.   **Tractive Effort:** 14 840 lbf.
Power Classification: Unclassified (0F).

1363   Didcot Railway Centre   Swindon 2377/1910

1366 CLASS   0-6-0PT

Built: 1934. Collett design for dock shunting. Used to work Weymouth Quay boat trains. 6 built. (1366–71).
Boiler Pressure: 165 lbf/sq. in.   **Weight:** 35.75 tons.
Wheel Diameter: 3' 8".   **Cylinders:** 16" x 20" (O).
Valve Gear: Stephenson. Slide valves.   **Tractive Effort:** 16 320 lbf.
Power Classification: Unclassified (1F).

1369   South Devon Railway   Swindon 1934

NORTH PEMBROKESHIRE & FISHGUARD RAILWAY   0-6-0ST

Built: 1878. Absorbed by GWR 1898. Sold to Gwendraeth Valley Railway in 1910. Absorbed by GWR again in 1923 but sold in March of that year to Kidwelly Tinplate Company.
Boiler Pressure: 140 lbf/sq. in.   **Weight:** 30.95 tons.
Wheel Diameter: 4' 0".   **Cylinders:** 16" x 22" (I).
Valve Gear: Stephenson. Slide valves.   **Tractive Effort:** 13 960 lbf.

GWR   *GVR*
1378   2   MARGARET   Scolton Manor Museum   FW 410/1878

1400 CLASS   0-4-2T

Built: 1932–36. Collett design. Push & Pull fitted. Locos renumbered in 1946. 75 built.(1400–74).
Boiler Pressure: 165 lbf/sq. in.   **Weight:** 41.3 tons.
Wheel Diameter: 5' 2", 3' 8".   **Cylinders:** 16" x 24" (I).
Valve Gear: Stephenson. Slide valves.   **Tractive Effort:** 13 900 lbf.
Power Class: Unclassified (1P).

1932 No.	*1946 No.*		
4820	1420	South Devon Railway	Swindon 1933
4842	1442	Tiverton Museum	Swindon 1935
4850	1450	Dean Forest Railway	Swindon 1935
4866	1466	Didcot Railway Centre	Swindon 1936

1500 CLASS   0-6-0PT

Built: 1949. Hawksworth design. 10 built (1500–09).
Boiler Pressure: 200 lbf/sq. in.   **Weight:** 58.2 tons.
Wheel Diameter: 4' 7½".   **Cylinders:** 17½" x 24" (O).
Valve Gear: Walschaerts. Piston valves.   **Tractive Effort:** 22 510 lbf.
Power Class: C (4F).   **Restriction:** Red.

1501   Severn Valley Railway   Swindon 1949

1600 CLASS   0-6-0PT

Built: 1949–55. Hawksworth design. 70 built (1600–69).
Boiler Pressure: 165 lbf/sq. in.   **Weight:** 41.6 tons.
Wheel Diameter: 4' 1½".   **Cylinders:** 16½" x 24" (I).
Valve Gear: Stephenson. Slide valves.   **Tractive Effort:** 18 510 lbf.
Power Class: A (2F).   **Restriction:** Uncoloured.

1638   Kent & East Sussex Railway   Swindon 1951

2251 CLASS   0-6-0

Built: 1930–48 Collett design. 120 built (2251–99, 2200–50, 3200–19).
Boiler Pressure: 200 lbf/sq. in. superheated.   **Weight–Loco:** 43.4 tons.
Wheel Diameter: 5' 2".   **–Tender:** 36.75 tons.
Cylinders: 17½" x 24" (I).   **Valve Gear:** Stephenson. Slide valves.
Tractive Effort: 20 150 lbf.
Power Class: B (3MT).   **Restriction:** Yellow.

3205	South Devon Railway	Swindon 1946

2301 CLASS   DEAN GOODS   0-6-0

Built: 1883–99. Dean design. 280 built (2301–2580).
Boiler Pressure: 180 lbf/sq. in. superheated.   **Weight–Loco:** 37 tons.
Wheel Diameter: 5' 2".   **–Tender:** 36.75 tons.
Cylinders: 17½" x 24" (I).   **Valve Gear:** Stephenson. Slide valves.
Tractive Effort: 18140 lbf.
Power Class: A (2MT).   **Restriction:** Uncoloured.

2516	Steam-Museum of the Great Western Railway (N)	Swindon 1557/1897

2800 CLASS   2-8-0

Built: 1903–19. Churchward design for heavy freight. 84 built (2800–83).
Boiler Pressure: 225 lbf/sq. in. superheated.   **Weight–Loco:** 75.5 tons.
Wheel Diameters: 3' 2", 4' 7½".   **–Tender:** 43.15 tons.
Cylinders: 18½" x 30" (O).   **Valve Gear:** Stephenson. Piston valves.
Tractive Effort: 35 380 lbf.
Power Class: E (8F).   **Restriction:** Blue.

2807	Gloucestershire-Warwickshire Railway	Swindon 2102/1905
2818	National Railway Museum	Swindon 2122/1905
2857	Severn Valley Railway	Swindon 2763/1918
2859	Llangollen Railway	Swindon 2765/1918
2861	Barry Island Railway	Swindon 2767/1918
2873	South Devon Railway	Swindon 2779/1918
2874	Pontypool & Blaenavon Railway	Swindon 2780/1918

2884 CLASS   2-8-0

Built: 1938–42. Collett development of 2800 Class with side window cabs. 81 built (2884–99, 3800–64).
Boiler Pressure: 225 lbf/sq. in. superheated.   **Weight–Loco:** 76.25 tons.
Wheel Diameters: 3' 2", 4' 7½".   **–Tender:** 43.15 tons.
Cylinders: 18½" x 30" (O).   **Valve Gear:** Stephenson. Piston valves.
Tractive Effort: 35 380 lbf.
Power Class: E (8F).   **Restriction:** Blue.

2885	Moor Street Station, Birmingham	Swindon 1938
3802	Llangollen Railway	Swindon 1938
3803	South Devon Railway	Swindon 1939
3814	North Yorkshire Moors Railway	Swindon 1940
3822	Didcot Railway Centre	Swindon 1940
3845	Rye Farm, Wishaw	Swindon 1942
3850	West Somerset Railway	Swindon 1942
3855	Pontypool & Blaenavon Railway	Swindon 1942
3862	Northampton & Lamport Railway	Swindon 1942

3200 CLASS   "DUKEDOG"   4-4-0

Built: Rebuilt 1936–39 by Collett using the frames of "Bulldogs" and the boilers of "Dukes". 30 built (9000–29).
Boiler Pressure: 180 lbf/sq. in.
Wheel Diameters: 3' 8", 5' 8".   **Weight–Loco:** 49 tons.
Cylinders: 18" x 26" (I).   **–Tender:** 40 tons.
Tractive Effort: 18 950 lbf.   **Valve Gear:** Stephenson. Slide valves.
Power Class: B.
     Restriction: Yellow.

3217–9017	"EARL OF BERKELEY"	Bluebell Railway	Swindon 1938

3700 CLASS   CITY   4-4-0

Built: 1903. Churchward design. Reputed to be the first loco to attain 100 m.p.h. when it hauled an "Ocean Mails" special from Plymouth to Paddington in 1904.
Boiler Pressure: 200 lbf/sq. in. superheated.   **Weight –Loco:** 55.3 tons.
Wheel Diameters: 3′ 2″, 6′ 8½″.   **–Tender:** 36.75 tons.
Cylinders: 18″ x 26″ (I).   **Valve Gear:** Stephenson. Piston valves.
Tractive Effort: 17 800 lbf.

BR   GWR
3440   3717   CITY OF TRURO   National Railway Museum   Swindon 2000/1903

4000 CLASS   STAR   4-6-0

Built: 1906–23. Churchward design for express passenger trains. 73 built (4000–72).
Boiler Pressure: 225 lbf/sq. in. superheated.   **Weight –Loco:** 75.6 tons.
Wheel Diameters: 3′ 2″, 6′ 8½″.   **–Tender:** 40 tons.
Cylinders: 15″ x 26″ (4).   **Tractive Effort:** 27 800 lbf.
Valve Gear: Inside Walschaerts. Rocking levers for outside valves. Piston valves.
Power Class: D (5P).   **Restriction:** Red.

4003   LODE STAR   National Railway Museum   Swindon 2231/1907

4073 CLASS   CASTLE   4-6-0

Built: 1923–50. Collett development of Star. 166 built (4073–99, 5000–5099, 7000–37). In addition one Pacific (111) and five Stars (4000/9/16/32/7) were rebuilt as Castles.
Boiler Pressure: 225 lbf/sq. in. superheated.   **Weight –Loco:** 79.85 tons.
Wheel Diameters.: 3′ 2″, 6′ 8½″.   **–Tender:** 46.7 tons.
Cylinders.: 16″ x 26″ (4).   **Tractive Effort:** 31630 lbf.
Valve Gear: Inside Walschaerts. Rocking levers for outside valves. Piston valves.
Power Class: D (7P).   **Restriction:** Red.

d–Rebuilt with double chimney.   x–Dual (air/vacuum) brakes.

4073	CAERPHILLY CASTLE	Steam-Museum of the Great Western Railway (N)	Swindon 1923
4079	PENDENNIS CASTLE	Didcot Railway Centre	Swindon 1924
5029 x	NUNNEY CASTLE	Old Oak Common Depot, London	Swindon 1934
5043 d	EARL OF MOUNT EDGCUMBE	Tyseley Locomotive Works, Birmingham	Swindon 1936
5051	EARL BATHURST	Didcot Railway Centre	Swindon 1936
5080	DEFIANT	Buckinghamshire Railway Centre	Swindon 1939
7027	THORNBURY CASTLE	The Railway Age, Crewe	Swindon 1949
7029 d	CLUN CASTLE	Tyseley Locomotive Works, Birmingham	Swindon 1950

5043 was named BARBURY CASTLE to 09/37.
5051 was named DRYSLLWYN CASTLE to 08/37.
5080 was named OGMORE CASTLE to 01/41.

Note: 5080 is on loan from Tyseley Locomotive Works, Birmingham

4200 CLASS   2-8-0T

Built: 1910–23. Churchward design. 105 built (4201–99, 4200, 5200–4).
Boiler Pressure: : 200 lbf/sq. in. superheated.   **Weight:** 81.6 tons.
Wheel Diameters.: 3′ 2″, 4′ 7½″.   **Cylinders.:** 18½″ x 30″ (O).
Valve Gear: Stephenson. Piston valves.   **Tractive Effort:** 31 450 lbf.
Power Class: E (7F).   **Restriction:** Red.

4247	Bodmin Steam Railway	Swindon 2637/1916
4248	Steam-Museum of the Great Western Railway	Swindon 2638/1916
4253	Pontypool & Blaenavon Railway	Swindon 2643/1917
4270	Gloucestershire-Warwickshire Railway	Swindon 2850/1919
4277	Swindon & Cricklade Railway	Swindon 2857/1920

◀ 4500 Class 2-6-2T No. 5552 leaves Bodmin on the delightful Bodmin & Wenford Railway on 5 June 2004. **Alan Barnes**

▼ 5101 Class 2-6-2T No. 5164 crosses Oldbury Viaduct on the Severn Valley Railway with the 15.35 Bridgnorth–Kidderminster on 17 April 2006.
Hugh Ballantyne

4300 CLASS 2-6-0

Built: 1911–32. Churchward design. 342 built (4300–99 (renumbered from 8300–99 between 1944 and 1948), 6300–99, 7300–21, 7322–41 (renumbered from 9300–19 between 1956 and 1959).
Boiler Pressure: 200 lbf/sq. in. superheated. **Weight –Loco:** 62 tons.
Wheel Diameters.: 3′ 2″, 5′ 8″. —**Tender:** 40 tons.
Cylinders.: 18½″ x 30 (O). **Valve Gear:** Stephenson. Piston valves.
Tractive Effort: 25 670 lbf.
Power Class: D (4MT). **Restriction:** Blue.

8322–5322	Didcot Railway Centre	Swindon 1917
9303*–7325	Steam – Museum of the Great Western Railway	Swindon 1932

4500 CLASS 2-6-2T

Built: 1906–24. Churchward design. (§Built 1927–29. Collett development with larger tanks). 175 built (4500–99, 5500–74).
Boiler Pressure: 200 lbf/sq. in. superheated. **Weight:** 57.9 tons (61 tons§).
Wheel Diameters: 3′ 2″, 4′ 7½″, 3′ 2″. **Cylinders:** 17″ x 24″ (O).
Valve Gear: Stephenson. Piston valves. **Tractive Effort:** 21 250 lbf.
Power Class: C (4MT). **Restriction:** Yellow.

4555	"WARRIOR"	Paignton & Dartmouth Railway	Swindon 1924
4561		West Somerset Railway	Swindon 1924
4566		Severn Valley Railway	Swindon 1924
4588§	"TROJAN"	Paignton & Dartmouth Railway	Swindon 1927
5521§		Dean Forest Railway	Swindon 1927
5526§		South Devon Railway	Swindon 1928
5532§		Llangollen Railway	Swindon 1928
5538§		Dean Forest Railway	Swindon 1928
5539§		Llangollen Railway	Swindon 1928
5541§		Dean Forest Railway	Swindon 1928
5542§		West Somerset Railway	Swindon 1928
5552§		Bodmin Steam Railway	Swindon 1928
5553§		West Somerset Railway	Swindon 1928
5572§		Didcot Railway Centre	Swindon 1929

4900 CLASS HALL 4-6-0

Built: 1928–43. Collett development of Churchward 'Saint' class. 259 in class. (4900 rebuilt from Saint) 4901–99, 5900–99, 6900–58 built as Halls).
Boiler Pressure: 225 lbf/sq. in. superheated. **Weight –Loco:** 75 tons.
Wheel Diameters: 3′ 2″, 6′ 0″. —**Tender:** 46.7 tons.
Cylinders: 18½″ x 30″ (O). **Valve Gear:** Stephenson. Piston valves.
Tractive Effort: 27 270 lbf.
Power Class: D (5MT). **Restriction:** Red.

4920	DUMBLETON HALL	South Devon Railway	Swindon 1929
4930	HAGLEY HALL	Severn Valley Railway	Swindon 1929
4936	KINLET HALL	Tyseley Locomotive Works, Birmingham	Swindon 1929
4953	PITCHFORD HALL	Tyseley Locomotive Works, Birmingham	Swindon 1929
4965	ROOD ASHTON HALL	Tyseley Locomotive Works, Birmingham	Swindon 1930
4979	WOOTTON HALL	Appleby Heritage Centre	Swindon 1930
5900	HINDERTON HALL	Didcot Railway Centre	Swindon 1931
5952	COGAN HALL	Cambrian Railway Trust, Llynclys	Swindon 1935
5967	BICKMARSH HALL	Pontypool & Blaenavon Railway	Swindon 1937
5972	OLTON HALL	West Coast Railway Company, Carnforth	Swindon 1937

Notes: 4965 previously carried the identity 4983 ALBERT HALL.
5972 at present masquerades as "HOGWARTS CASTLE" for use in "Harry Potter" films.

5101 CLASS   2-6-2T

Built: 1929–49. Collett development of Churchward 3100 class. 180 built (5101–99, 4100–79).
Boiler Pressure: 200 lbf/sq. in. superheated.   **Weight**: 78.45 tons.
Wheel Diameters: 3' 2", 5' 8", 3' 8".   **Cylinders**: 18" x 30" (O).
Valve Gear: Stephenson. Piston valves.   **Tractive Effort**: 24 300 lbf.
Power Class: D (4MT).   **Restriction**: Yellow.

4110	Tyseley Locomotive Works, Birmingham	Swindon 1936
4115	Barry Island Railway	Swindon 1936
4121	Tyseley Locomotive Works, Birmingham	Swindon 1937
4141	Great Central Railway	Swindon 1946
4144	Didcot Railway Centre	Swindon 1946
4150	Severn Valley Railway	Swindon 1947
4160	West Somerset Railway	Swindon 1948
5164	Severn Valley Railway	Swindon 1930
5199	Llangollen Railway	Swindon 1934

Note: 5193 has been rebuilt by the West Somerset Railway as 4300 Class 2-6-0 tender locomotive No. 9351. See section 1.1.1.

5205 CLASS   2-8-0T

Built: 1923–25/40. Collett development of 4200 class. 60 built (5205–64).
Boiler Pressure: 200 lbf/sq. in. superheated.   **Weight**: 82.1 tons.
Wheel Diameters: 3' 2", 4' 7½".   **Cylinders**: 19" x 30" (O).
Valve Gear: Stephenson. Piston valves.   **Tractive Effort**: 33 170 lbf.
Power Class: E (8F).   **Restriction**: Red.

5224		West Somerset Railway	Swindon 1925
5227		Barry Island Railway	Swindon 1924
5239	"GOLIATH"	Paignton & Dartmouth Railway	Swindon 1924

5600 CLASS   0-6-2T

Built: 1924–28. Collett design. 200 built (5600–99, 6600–99).
Boiler Pressure: 200 lbf/sq. in. superheated.   **Weight**: 68 tons.
Wheel Diameters: 4' 7½", 3' 8".   **Cylinders**: 18" x 26" (I).
Valve Gear: Stephenson. Piston valves.   **Tractive Effort**: 25 800 lbf.
Power Class: D (5MT).   **Restriction**: Red.

5619	Telford Steam Railway	Swindon 1925
5637	East Somerset Railway	Swindon 1925
5643	Lakeside & Haverthwaite Railway	Swindon 1925
5668	Pontypool & Blaenavon Railway	Swindon 1926
6619	North Yorkshire Moors Railway	Swindon 1928
6634	The Railway Age, Crewe	Swindon 1928
6686	Barry Island Railway	AW 974/1928
6695	Swanage Railway	AW 983/1928
6697	Didcot Railway Centre	AW 985/1928

Note: 5637 is on loan from the Swindon & Cricklade Railway.

5700 CLASS   0-6-0PT

Built: 1929–49. Collett design. The standard GWR shunter. 863 built (5700–99, 6700–79, 7700–99, 8700–99, 3700–99, 3600–99, 4600–99, 9600–82, 9700–9799. Six of the preserved examples saw use with London Transport following withdrawal by British Railways.
Boiler Pressure: 200 lbf/sq. in.   **Weight**: 47.5 (49§) tons.
Wheel Diameter: 4' 7½"   **Cylinders**: 17½" x 24" (I).
Valve Gear: Stephenson. Slide valves..   **Tractive Effort**: 22 510 lbf.
Power Class: C (4F).   **Restriction**: Blue (Yellow from 1950).

GWR	LTE		
3650		Didcot Railway Centre	Swindon 1939
3738		Didcot Railway Centre	Swindon 1937
4612		Bodmin Steam Railway	Swindon 1942
5764	L95	Severn Valley Railway	Swindon 1929
5775	L89	Keighley & Worth Valley Railway	Swindon 1929

5786	L92	South Devon Railway	Swindon 1930
7714		Severn Valley Railway	KS 4449/1930
7715	L99	Buckinghamshire Railway Centre	KS 4450/1930
7752	L94	Tyseley Locomotive Works, Birmingham	NBL 24040/1930
7754		Llangollen Railway	NBL 24042/1930
7760	L90	Tyseley Locomotive Works, Birmingham	NBL 24048/1930
9600§		Tyseley Locomotive Works, Birmingham	Swindon 1945
9629§		Pontypool & Blaenavon Railway	Swindon 1946
9642§		Gloucestershire-Warwickshire Railway	Swindon 1946
9681§		Dean Forest Railway	Swindon 1949
9682§		North Norfolk Railway	Swindon 1949

Note: 9682 is on loan from the GWR Preservation Group.

6000 CLASS KING 4-6-0

Built: 1927–30. Collett design. 31 built.
Boiler Pressure: 250 lbf/sq. in. superheated.
Wheel Diameters: 3′ 0″, 6′ 6″.
Cylinders: 16¼″ x 28″ (4).
Valve Gear: Inside Walschaerts. Piston valves. Rocking levers for outside valves.
Power Class: Special (8P).
Weight –Loco: 89 tons.
              –Tender: 46.7 tons.
Tractive Effort: 40 290 lbf.
Restriction: Double Red.
x–Dual (air/vacuum) brakes.

6000	KING GEORGE V	Steam – Museum of the Great Western Railway (N)	Swindon 1927
6023	KING EDWARD II	Didcot Railway Centre	Swindon 1930
6024 x	KING EDWARD I	Didcot Railway Centre	Swindon 1930

6100 CLASS 2-6-2T

Built: 1931–35. Collett development of 5100. 70 built (6100–69).
Boiler Pressure: 225 lbf/sq. in. superheated.
Wheel Diameters: 3′ 2″, 5′ 8″, 3′ 8″.
Valve Gear: Stephenson. Piston valves.
Power Class: D (5MT).
Weight: 78.45 tons.
Cylinders: 18″ x 30″ (O).
Tractive Effort: 27 340 lbf.
Restriction: Blue.

6106	Didcot Railway Centre	Swindon 1931

6400 CLASS 0-6-0PT

Built: 1932–37. Collett design. Push & Pull fitted. 40 built (6400–39).
Boiler Pressure: 165 lbf/sq. in.
Wheel Diameter: 4′ 7½″.
Valve Gear: Stephenson. Slide valves..
Power Class: A (2P).
Weight: 45.6 tons.
Cylinders: 16½″ x 24″ (I).
Tractive Effort: 16 510 lbf.
Restriction: Yellow.

6412		West Somerset Railway	Swindon 1934
6430		Llangollen Railway	Swindon 1937
6435	"AJAX"	Paignton & Dartmouth Railway	Swindon 1937

6959 CLASS MODIFIED HALL 4-6-0

Built: 1944–49. Hawksworth development of 'Hall'. 71 built (6959–99, 7900–29).
Boiler Pressure: 225 lbf/sq. in. superheated.
Wheel Diameters: 3′ 2″, 6′ 0″.
Cylinders: 18½″ x 30″ (O).
Tractive Effort: 27 270 lbf.
Power Class: D (5MT).
Weight –Loco: 75.8 tons.
              –Tender: 47.3 tons.
Valve Gear: Stephenson. Piston valves.
Restriction: Blue.

6960	RAVENINGHAM HALL	Gloucestershire-Warwickshire Railway	Swindon 1944
6984	OWSDEN HALL	Gloucestershire-Warwickshire Railway	Swindon 1948
6989	WIGHTWICK HALL	Buckinghamshire Railway Centre	Swindon 1948
6990	WITHERSLACK HALL	Great Central Railway	Swindon 1948
6998	BURTON AGNES HALL	Didcot Railway Centre	Swindon 1949
7903	FOREMARKE HALL	Gloucestershire-Warwickshire Railway	Swindon 1949

Note: 7903 is on loan from the Swindon & Cricklade Railway.

▲ The Great Central Railway's 13.15 service from Loughborough Central to Leicester hauled by No. 7821 DITCHEAT MANOR passing Kinchley Lane on 26 June 2005.
Hugh Ballantyne

◀ 5600 Class 0-6-2T No. 5637, masquerading as No. 5689, at Cranmore on the East Somerset Railway with a Russ Hillier photo charter on 17 January 2004. **Alan Barnes**

7200 CLASS   2-8-2T
Built: 1934–50. Collett rebuilds of 4200 and 5205 class 2-8-0Ts. 54 built (7200–53).
Boiler Pressure: 200 lbf/sq. in. superheated.   **Weight:** 92.6 tons.
Wheel Diameters: 3' 2", 4' 7½", 3' 8".   **Cylinders:** 19" x 30" (O).
Valve Gear: Stephenson. Piston valves.   **Tractive Effort:** 33 170 lbf
Power Class: E (8F).   **Restriction:** Blue.

7200	(rebuilt from 5277)	Buckinghamshire Railway Centre	Swindon 1930 reb. 1934
7202	(rebuilt from 5275)	Didcot Railway Centre	Swindon 1930 reb. 1934
7229	(rebuilt from 5264)	East Lancashire Railway	Swindon 1926 reb. 1935

7800 CLASS   MANOR   4-6-0
Built: 1938–50. Collett design for secondary main lines. 30 built (7800–29).
Boiler Pressure: 225 lbf/sq. in. superheated.   **Weight–Loco:** 68.9 tons.
Wheel Diameters: 3' 0", 5' 8".   **–Tender:** 40 tons.
Cylinders: 18" x 30" (O).   **Valve Gear:** Stephenson. Piston valves.
Tractive Effort: 27 340 lbf.
Power Class: D (5MT).   **Restriction:** Blue.

7802	BRADLEY MANOR	Severn Valley Railway	Swindon 1938
7808	COOKHAM MANOR	Didcot Railway Centre	Swindon 1938
7812	ERLESTOKE MANOR	Severn Valley Railway	Swindon 1939
7819	HINTON MANOR	Designer Outlet Village, Swindon	Swindon 1939
7820	DINMORE MANOR	West Somerset Railway	Swindon 1950
7821	DITCHEAT MANOR	Churnet Valley Railway	Swindon 1950
7822	FOXCOTE MANOR	Llangollen Railway	Swindon 1950
7827	LYDHAM MANOR	Paignton & Dartmouth Railway	Swindon 1950
7828	ODNEY MANOR	West Somerset Railway	Swindon 1950

Note: 7821 is on loan from the Cambrian Railway Trust, Llynclys.

9400 CLASS   0-6-0PT
Built: 1947–56. Hawksworth design. 210 built (9400–99, 8400–99, 3400–09).
Boiler Pressure: 200 lbf/sq. in. (*Superheated)   **Weight:** 55.35 tons.
Wheel Diameter: 4' 7½".   **Cylinders:** 17½" x 24" (I).
Valve Gear: Stephenson. Slide valves.   **Tractive Effort:** 22 510 lbf.
Power Class: C (4F).   **Restriction:** Red.

9400*	Steam – Museum of the Great Western Railway (N)	Swindon 1947
9466	Buckinghamshire Railway Centre	RSH 7617/1952

BURRY PORT & GWENDRAETH VALLEY RAILWAY   0-6-0ST
Built: 1900.   **Weight:** 29 tons.
Wheel Diameters: 3' 6".   **Cylinders:** 14" x 20"(O)
Note: This loco was supplied new to the BPGVR and was sold into industrial service in 1914.

2	PONTYBEREM	Didcot Railway Centre	AE 1421/1900

SANDY & POTTON RAILWAY   0-4-0WT
Built: 1857. The Sandy & Potton Railway became part of the LNWR and the loco worked on the Cromford & High Peak Railway from 1863–1878. The loco was sold to the Wantage Tramway in 1878.
Boiler Pressure: 120 lbf/sq. in.   **Weight:** 15 tons.
Wheel Diameter: 3' 0"   **Cylinders:** 9" x 12" (O)
Tractive Effort: 5510 lbf.

SPR	LNWR	WT		
SHANNON	1863	5	Didcot Railway Centre (N)	GE 1857

SOUTH DEVON RAILWAY   0-4-0WT
Built: 1868. Vertical boilered locomotive.   **Gauge:** 7' 0¼".
Wheel Diameter: 3' 0"   **Cylinders:** 9" x 12" (V).

GWR	SDR		
2180	151	TINY	South Devon Railway (N)	Sara 1868

1.1.1. NEW BUILD PROJECTS

There are various Great Western designs which are being produced using parts from other locos to a greater or lesser extent. These are detailed here. NB: Technical details refer to the original locos and the new builds may differ from these.

1000 CLASS — COUNTY — 4-6-0

Original Class Built: 1945–47. 30 built. 1000–29. The new build loco uses the frames from 7927 WILLINGTON HALL and assumes the identity of a former member of the class. The boiler is rebuilt from Stanier Class 8F 2-8-0 48518.
Boiler Pressure: 250 lbf/sq. in. superheated. **Weight –Loco:** 76.85 tons.
Wheel Diameters: 3' 0", 6'3". **–Tender:** 49.00 tons.
Cylinders: 18½" x 30" (O). **Valve Gear:** Stephenson. Piston valves.
Tractive Effort: 29 090 lbf.
Power Class: D (6MT). **Restriction:** Red.

1014 COUNTY OF GLAMORGAN   Didcot Railway Centre   Under construction

2900 CLASS — SAINT — 4-6-0

Original Class Built: 1902–13. 77 built. 2910–55, 2971–90, 2998. The new build loco is rebuilt from 4942 MAINDY HALL.
Boiler Pressure: 225 lbf/sq. in. superheated. **Weight –Loco:** 72.0 tons.
Wheel Diameters: 3' 2", 6' 8½". **–Tender:** 43.15 tons.
Cylinders: 18½" x 30" (O). **Valve Gear:** Stephenson.
Tractive Effort: 24 390 lbf.
Power Class: D (4P). **Restriction:** Red.

2999 LADY OF LEGEND   Didcot Railway Centre   Under construction

4300 CLASS — 2-6-0

Rebuilt from 5101 Class 2-6-2T No. 5193. Dimensions differ from standard 4300 Class locos.
Boiler Pressure: 225 lbf/sq. in. superheated. **Weight –Loco:** 63.85 tons.
Wheel Diameters.: 3' 2", 5' 8". **–Tender:** 40 tons.
Cylinders.: 18" x 30 (O). **Valve Gear:** Stephenson. Piston valves.
Tractive Effort: 28 880 lbf.
Power Class: D (4MT). **Restriction:** Blue.

9351   West Somerset Railway Swindon 1934 rebuilt WSR 2006

6800 CLASS — GRANGE — 4-6-0

Original Class Built: 1936–39. 80 built. 6800–79. The new build loco uses the boiler of 7927 WILLINGTON HALL.
Boiler Pressure: 225 lbf/sq. in. superheated. **Weight –Loco:** 74.00 tons.
Wheel Diameters: 3' 0", 5' 8". **–Tender:** 40.00 tons.
Cylinders: 18½" x 30" (O). **Valve Gear:** Stephenson. Piston valves.
Tractive Effort: 28 875 lbf.
Power Class: D (5MT). **Restriction:** Red.

6880 BETTON GRANGE   Llangollen Railway   Under construction

SOUTHERN

1.2. SOUTHERN RAILWAY AND CONSTITUENT COMPANIES' STEAM LOCOMOTIVES

GENERAL
The Southern Railway (SR) was an amalgamation of the London, Brighton & South Coast Railway (LBSCR), the London & South Western Railway (LSWR) and the South Eastern & Chatham Railway (SECR). The latter was formed in 1898 by the amalgamation of the South Eastern Railway (SER) and London, Chatham & Dover Railway (LCDR).

Locomotive Numbering System
On formation of the SR in 1924, all locomotives (including new builds) were given a prefix letter to denote the works which maintained them as follows:

A   Ashford Works. All former SECR locomotives plus some D1, L1, U1.
B   Brighton Works. All former LBSCR locomotives plus some D1.
E   Eastleigh Works. All former LSWR locomotives plus LN, V, Z.

In 1931 locomotives were renumbered. "E" prefix locomotives merely lost the prefix (except locomotives with an "0" in front of the number to which 3000 was added). "A" prefix locomotives had 1000 added and "B" prefix locomotives had 2000 added, e.g. B636 became 2636, E0298 became 3298.

In 1941 Bulleid developed a most curious numbering system for his new locomotives. This consisted of two numbers representing the numbers of leading and trailing axles respectively followed by a letter denoting the number of the driving axles. This was followed by the locomotive serial number. The first pacific was therefore 21C1, and the first Q1 0-6-0 was C1.

In 1948 British Railways added 30000 to all numbers, but the 3xxx series (formerly 0xxx series) were totally renumbered. The Q1s became 33xxx, the MNs 35xxx and the WCs 34xxx. Isle of Wight locomotives had their own number series, denoted by a "W" prefix. This indicated Ryde Works maintenance and was carried until the end of steam.

In the section which follows, locomotives are listed generally in order of BR numbers. Three old locomotives which were withdrawn before nationalisation are listed at the end of the section.

Classification
The LBSCR originally classified locomotive classes by a letter which denoted the use of the class. A further development was to add a number, to identify different classes of similar use. A rebuild was signified by an "X" suffix. In its latter years, new classes of different wheel arrangement were given different letters. The SECR gave each class a letter. A number after the letter signified either a new class which was a modification of the original or a rebuild. The SR perpetuated this system. The LSWR had an odd system based on the works order number for the first locomotive of the class to be built. These went A1, B1......Z1, A2......Z2, A3......etc. and did not only apply to locomotives. Locomotives bought from outside contractors were classified by the first number to be delivered, e.g. "0298 Class".

▲ Adams LSWR 0-4-0T Class B4 No. 30096 at Horsted Keynes on the Bluebell Railway on 29 August 2004.
Steve Widdowson

▶ Beattie LSWR well tank No. 30587 (0298 Class) with a demonstration night freight near Bodmin on 29 November 2002.
Alan Barnes

▲ Class N15 (King Arthur) 4-6-0 No. 30777 SIR LAMIEL at Slitting Mill Crossing (north of Barrow Hill) with a Birmingham–York special on 7 August 2006. **Peter Fox**

▼ Class LN 4-6-0 No. 850 LORD NELSON passes Dunball as it heads away from Bridgwater with the Kingfisher Railtours "Trafalgar Express" from Minehead to Eastleigh on 31 March 2007.
Rail Photoprints

CLASS O2 — 0-4-4T

Built: 1889–91. Adams LSWR design.
Boiler Pressure: 160 lbf/sq. in.
Wheel Diameters: 4' 10"
Valve Gear: Stephenson. Slide valves.
BR Power Classification: 1P
Weight: 48.4 tons.
Cylinders: 17" x 24" (I).
Tractive Effort: 17 235 lbf.
Air braked.

BR	SR	LSWR			
W24	E209–W24	209 CALBOURNE	Isle of Wight Steam Railway		Nine Elms 341/1891

CLASS M7 — 0-4-4T

Built: 1897–1911. Drummond LSWR design. 105 built.
Boiler Pressure: 175 lbf/sq. in.
Wheel Diameters: 5' 7", 3' 7".
Valve Gear: Stephenson. Slide valves.
BR Power Classification: 2P.
Weight: 60.15 tons.
Cylinders: 18½" x 26" (I).
Tractive Effort: 19 760 lbf.

30053 was push and pull fitted and air braked.

BR	SR	LSWR		
30053	E53–53	53	Swanage Railway	Nine Elms 1905
30245	E245–245	245	National Railway Museum	Nine Elms 501/1897

CLASS USA — 0-6-0T

Built: 1942–43 by Vulcan Works, Wilkes-Barre, PA, USA for US Army Transportation Corps. 93 built (a further 289 were built by other builders). Thirteen were sold to SR in 1947 becoming 62–74.
Boiler Pressure: 210 lbf/sq. in.
Wheel Diameter: 4' 6"
Valve Gear: Walschaerts. Piston valves.
BR Power Classification: 3F.
Weight: 46.5 tons.
Cylinders: 16½" x 24" (O).
Tractive Effort: 21 600 lbf.

BR	SR	USATC	Present		
30064	64	1959		Bluebell Railway	VIW 4432/1943
30065	65	1968	"MAUNSELL"	Kent & East Sussex Railway	VIW 4441/1943
30070	70	1960	"WAINWRIGHT"	Kent & East Sussex Railway	VIW 4433/1943
30072	72	1973		Keighley & Worth Valley Railway	VIW 4446/1943

Note: 30065 carried DS237 and 30070 carried DS238 from 1963.

CLASS B4 — 0-4-0T

Built: 1891–1909. Adams LSWR design for dock shunting (25 built).
Boiler Pressure: 140 lbf/sq. in.
Wheel Diameter: 3' 9¾".
Valve Gear: Stephenson. Slide valves.
BR Power Classification: 1F.
Weight: 33.45 tons.
Cylinders: 16" x 22" (O).
Tractive Effort: 14 650 lbf.

BR	SR			
30096	E96–96	NORMANDY	Bluebell Railway	Nine Elms 396/1893
30102	E102–102	GRANVILLE	Bressingham Steam Museum	Nine Elms 406/1893

CLASS T9 — 4-4-0

Built: 1889–1924. Drummond LSWR express passenger design. 66 built.
Boiler Pressure: 175 lbf/sq. in. Superheated
Wheel Diameters: 3' 7", 6' 7".
Cylinders: 19" x 26" (I).
Tractive Effort: 17 670 lbf.
Weight –Loco: 51.8 tons.
    –Tender: 44.85 tons.
Valve Gear: Stephenson. Slide valves.
BR Power Classification: 3P.

BR	SR	LSWR		
30120	E120–120	120	Bluebell Railway (N)	Nine Elms 572/1899

CLASS S15 (URIE) 4-6-0
Built: 1920–21. Urie LSWR design. 20 built.
Boiler Pressure: 180 lbf/sq. in. superheated.
Wheel Diameters: 3′ 7″, 5′ 7″.
Cylinders: 21″ x 28″ (O).
Tractive Effort: 28 200 lbf.
Weight –Loco: 79.8 tons.
      –Tender: 57.8 tons.
Valve Gear: Walschaerts. Piston valves.
BR Power Classification: 6F.

BR	SR	LSWR		
30499	E499–499	499	East Lancashire Railway	Eastleigh 1920
30506	E506–506	506	Mid Hants Railway	Eastleigh 1920

CLASS Q 0-6-0
Built: 1938–39. Maunsell SR design. 20 built (30530–49).
Boiler Pressure: 200 lbf/sq. in. superheated.
Wheel Diameters: 5′ 1″.
Cylinders: 19″ x 26″ (I).
Tractive Effort: 26 160 lbf.
Weight –Loco: 49.5 tons.
      –Tender: 40.5 tons.
Valve Gear: Stephenson. Piston valves.
BR Power Classification: 4F.

BR	SR		
30541	541	Bluebell Railway	Eastleigh 1939

0415 CLASS 4-4-2T
Built: 1882–85. Adams LSWR design. 72 built.
Boiler Pressure: 160 lbf/sq. in.
Wheel Diameters: 3′ 0″, 5′ 7″, 3′ 0″
Valve Gear: Stephenson. Slide valves.
BR Power Classification: 1P.
Weight: 55.25 tons.
Cylinders: 17½″ x 24″ (O).
Tractive Effort: 14 920 lbf.

BR	SR	LSWR		
30583	E0488–3488	488	Bluebell Railway	N 3209/1885

0298 CLASS 2-4-0WT
Built: 1863–75. W. G. Beattie LSWR design. Last used on the Wenfordbridge branch in Cornwall. 85 built. Survivors reboilered in 1921.
Boiler Pressure: 160 lbf/sq. in.
Wheel Diameters: 3′ 7¾″, 5′ 7″.
Valve Gear: Allan.
BR Power Classification: 0P.
Weight: 35.75 (36.3§) tons.
Cylinders: 16½″ x 22″ (O).
Tractive Effort: 12 160 lbf.

BR	SR	LSWR		
30585§	E0314–3314	0314	Buckinghamshire Railway Centre	BP 1414/1874 reb Elh 1921
30587	E0298–3298	0298	Bodmin Steam Railway (N)	BP 1412/1874 reb Elh 1921

CLASS N15   KING ARTHUR 4-6-0
Built: 1925–27. Maunsell SR development of Urie LSWR design. 54 built (30448–457, 30763–806).
Boiler Pressure: 200 lbf/sq. in. superheated.
Wheel Diameters: 3′ 7″, 6′ 7″.
Cylinders: 20½″ x 28″ (O).
Tractive Effort: 25 320 lbf.
Weight –Loco: 80.7 tons.
      –Tender: 57.5 tons.
Valve Gear: Walschaerts. Piston valves.
BR Power Classification: 5P.

BR	SR			
30777	777	SIR LAMIEL	Great Central Railway (N)	NBL 23223/1925

26

▲ Class A1X "Terrier" 0-6-0T FENCHURCH carrying No. 672 at Sheffield Park on the Bluebell Railway on 29 August 2004. **Steve Widdowson**

▼ Billinton LBSCR Class E4 0-6-2T No. 32473 at Kingscote on the Bluebell Railway on 14 October 2005. **Chris Wilson**

CLASS S15 (MAUNSELL)   4-6-0
Built: 1927–36. Maunsell SR development of Urie LSWR design.
Boiler Pressure: 200 lbf/sq. in. superheated.   **Weight –Loco:** 80.7 (79.25*) tons.
Wheel Diameters: 3′ 7″, 5′ 7″.   **–Tender:** 56.4 tons.
Cylinders: 20½″ x 28″ (O).   **Valve Gear:** Walschaerts. Piston valves.
Tractive Effort: 29 860 lbf.   **BR Power Classification:** 6F.

BR	SR			
30825	825		North Yorkshire Moors Railway	Eastleigh 1927
30828	828	"HARRY A FRITH"	Mid Hants Railway	Eastleigh 1927
30830	830		North Yorkshire Moors Railway	Eastleigh 1927
30847*	847		Bluebell Railway	Eastleigh 1936

Note: 30825 has been restored using a substantial number of components from 30841.

CLASS LN   LORD NELSON   4-6-0
Built: 1926–29. Maunsell SR design. 16 built (30850–65).
Boiler Pressure: 220 lbf/sq. in. superheated.   **Weight –Loco:** 83.5 tons.
Wheel Diameters: 3′ 1″, 6′ 7″.   **–Tender:** 57.95 tons.
Cylinders: 16½″ x 26″ (4).   **Valve Gear:** Walschaerts. Piston valves.
Tractive Effort: 33 510 lbf.   **BR Power Classification:** 7P.
Dual (air/vacuum) brakes.

BR	SR			
30850	E850–850 LORD NELSON		National Railway Museum	Eastleigh 1926

CLASS V   SCHOOLS   4-4-0
Built: 1930–35. Maunsell SR design. 40 built (30900–39).
Boiler Pressure: 220 lbf/sq. in. superheated.   **Weight –Loco:** 67.1 tons.
Wheel Diameters: 3′ 1″, 6′ 7″.   **–Tender:** 42.4 tons.
Cylinders: 16½″ x 26″ (3).   **Valve Gear:** Walschaerts. Piston valves.
Tractive Effort: 25 130 lbf.   **BR Power Classification:** 5P.

BR	SR			
30925	925	CHELTENHAM	National Railway Museum	Eastleigh 1934
30926	926	REPTON	North Yorkshire Moors Railway	Eastleigh 1934
30928	928	STOWE	Bluebell Railway	Eastleigh 1934

CLASS P   0-6-0T
Built: 1909–10. Wainwright SECR design. 8 built.
Boiler Pressure: 160 lbf/sq. in.   **Weight:** 28.5 tons.
Wheel Diameters: 3′ 9″.   **Cylinders:** 12″ x 18″ (I).
Valve Gear: Stephenson. Slide valves.   **Tractive Effort:** 7830 lbf.
BR Power Classification: 0F.   Push & pull fitted.

BR	SR	SECR		
31027	A 27–1027	27	Bluebell Railway	Ashford 1910
31178	A178–1178	178	Bluebell Railway	Ashford 1910
31323	A323–1323	323	Bluebell Railway	Ashford 1910
31556	A556–1556	753	Kent & East Sussex Railway	Ashford 1909

CLASS O1   0-6-0
Built: 1903–15. Wainwright SECR design. 66 built. 59 were rebuilt out of 122 "O" class.
Boiler Pressure: 150 lbf/sq. in.   **Weight –Loco:** 41.05 tons.
Wheel Diameter: 5′ 1″.   **–Tender:** 28.20 tons.
Cylinders: 18″ x 26″ (I).   **Valve Gear:** Stephenson. Slide valves.
Tractive Effort: 17 610 lbf.   **BR Power Classification:** 1F.

BR	SR	SECR		
31065	A65–1065	65	Bluebell Railway	Ashford 1896 reb. 1908

CLASS H 0-4-4T
Built: 1904–15. Wainwright SECR design. 66 built.
Boiler Pressure: 160 lbf/sq. in.
Wheel Diameters: 5′ 6″, 3′ 7″.
Valve Gear: Stephenson. Slide valves.
BR Power Classification: 1P.
Weight: 54.4 tons.
Cylinders: 18″ x 26″ (I).
Tractive Effort: 17 360 lbf.
Push and Pull fitted. Air brakes.

BR	SR	SECR		
31263	A263–1263	263	Bluebell Railway	Ashford 1905

CLASS C 0-6-0
Built: 1900–08. Wainwright SECR design. 109 built.
Boiler Pressure: 160 lbf/sq. in.
Wheel Diameter: 5′ 2″.
Cylinders: 18½″ x 26″ (I).
Tractive Effort: 19 520 lbf.
Weight –Loco: 43.8 tons.
           –Tender: 38.25 tons.
Valve Gear: Stephenson. Slide valves.
BR Power Classification: 2F.

BR	SR	SECR		
31592–DS239	A592–1592	592	Bluebell Railway	Longhedge 1902

CLASS U 2-6-0
Built: 1928–31. Maunsell SR design. 50 built (31610–39, 31790–809). 31790–809 were converted from class K (River Class) 2-6-4Ts.
Boiler Pressure: 200 lbf/sq. in. superheated.
Wheel Diameters: 3′ 1″, 6′ 0″.
Cylinders: 19″ x 28″ (O).
Tractive Effort: 23 870 lbf.
Weight –Loco: 61.9 (62.55§) tons.
           –Tender: 42.4 tons.
Valve Gear: Walschaerts. Piston valves.
BR Power Classification: 4MT.

* Formerly class K 2-6-4T A806 RIVER TORRIDGE built Ashford 1926.

BR	SR		
31618	A618–1618	Bluebell Railway	Brighton 1928
31625	A625–1625	Mid Hants Railway	Ashford 1929
31638	A638–1638	Bluebell Railway	Ashford 1931
31806*	A806–1806	Mid Hants Railway	Brighton 1928

CLASS D 4-4-0
Built: 1901–07. Wainwright SECR design. 51 built.
Boiler Pressure: 175 lbf/sq. in.
Wheel Diameters: 3′ 7″, 6′ 8″.
Cylinders: 19¼″ x 26″ (I).
Tractive Effort: 17 910 lbf.
Weight –Loco: 50 tons.
           –Tender: 39.1 tons.
Valve Gear: Stephenson. Slide valves.
BR Power Classification: 1P.

BR	SR	SECR		
31737	A737–1737	737	National Railway Museum	Ashford 1901

CLASS N 2-6-0
Built: 1917–34. Maunsell SECR design. Some built by SR. 80 built.
Boiler Pressure: 200 lbf/sq. in.
Wheel Diameters: 3′ 1″, 5′ 6″.
Cylinders: 19″ x 28″ (O).
Tractive Effort: 26 040 lbf.
Weight –Loco: 59.4 tons.
           –Tender: 39.25 tons.
Valve Gear: Walschaerts. Piston valves.
BR Power Classification: 4MT.

BR	SR	Present		
31874	A874–1874	"5 JAMES"	Mid Hants Railway	Woolwich Arsenal 1925

CLASS E1   0-6-0T
Built: 1874–83. Stroudley LBSCR design. 80 built.
Boiler Pressure: 160 lbf/sq. in.   **Weight:** 44.15 tons.
Wheel Diameter: 4' 6"   **Cylinders:** 17" x 24" (I).
Valve Gear: Stephenson. Slide valves.   **Tractive Effort:** 17 470 lbf.

BR	SR	LBSCR			
–	B110	110		East Somerset Railway	Brighton 1877

CLASS E4   0-6-2T
Built: 1897–1903. R Billinton LBSCR design. 120 built.
Boiler Pressure: 160 lbf/sq. in.   **Weight:** 56.75 tons.
Wheel Diameters: 5' 0", 4' 0".   **Cylinders:** 18" x 26" (I).
Valve Gear: Stephenson. Slide valves.   **Tractive Effort:** 19 090 lbf.
BR Power Classification: 2MT.

R	SR	LBSCR			
32473	B473–2473	473	BIRCH GROVE	Bluebell Railway	Brighton 1898

CLASSES A1 & A1X   "TERRIER"   0-6-0T
Built: 1872–80 as class A1§. Stroudley LBSCR design. Most rebuilt to A1X from 1911. 50 built.
Boiler Pressure: 150 lbf/sq. in.   **Weight:** 28.25 tons.
Wheel Diameters: 4' 0".   **Cylinders:** 14" (13"†, 12"*) x 20" (I).
Valve Gear: Stephenson. Slide valves.   **Tractive Effort:** 10 410 lbf (8890 lbf§†, 7650 lbf*).
BR Power Classification: 0P.

a air brakes,
d dual (air/vacuum) brakes.

BR	SR	LBSCR			
32636 d†	B636–2636	72	FENCHURCH	Bluebell Railway	Brighton 1872
32640 a	W11–2640	40	NEWPORT	Isle of Wight Steam Railway	Brighton 1878
32646 a	W2–W8	46–646	FRESHWATER	Isle of Wight Steam Railway	Brighton 1876
32650 d*	B650–W9	50–650	WHITECHAPEL	Spa Valley Railway	Brighton 1876
DS680 a*	A751–680S	54–654	WADDON	Canadian Railroad Historical Museum	Brighton 1875
32655	B655–2655	55–655	STEPNEY	Bluebell Railway	Brighton 1875
32662 a*	B66–2662	62–662	MARTELLO	Bressingham Steam Museum	Brighton 1875
32670		70	POPLAR	Kent & East Sussex Railway	Brighton 1872
32678 d	B678–W4–W14	78–678	KNOWLE	Kent & East Sussex Railway	Brighton 1880
	380S	82–682	BOXHILL	National Railway Museum	Brighton 1880

Notes:

32640 was also named BRIGHTON.
32646 was sold to the LSWR and became 734. It has also been named NEWINGTON.
32650 became 515S (departmental) and was named FISHBOURNE when on the Isle of Wight. It is now named "SUTTON".
DS680 was sold to the SECR and became their 75.
32678 was named BEMBRIDGE when on the Isle of Wight.
32640 was originally Isle of Wight Central Railway No. 11
32646 was originally Freshwater Yarmouth and Newport Railway No. 2.

CLASS Q1   0-6-0
Built: 1942. Bulleid SR "Austerity" design. 40 built (33001–40).
Boiler Pressure: 230 lbf/sq. in. superheated.   **Weight –Loco:** 51.25 tons.
                                               –Tender: 38 tons
Wheel Diameter: 5' 1".   **Valve Gear:** Stephenson. Slide valves.
Cylinders: 19" x 26" (I).   **BR Power Classification:** 5F.
Tractive Effort: 30 080 lbf

BR	SR			
33001	C1		National Railway Museum	Brighton 1942

CLASSES WC & BB   4-6-2
WEST COUNTRY and BATTLE OF BRITAIN

Built: 1945–51. Bulleid SR design with "air smoothed" casing, thermic syphons and Boxpox driving wheels. 110 built (34001–110). All built at Brighton.
Boiler Pressure: 250 lbf/sq. in. superheated.   **Weight –Loco:** 86 (91.65*) tons.
Wheel Diameters: 3' 1", 6' 2", 3' 1"   **–Tender:** 42.7, 47.9 or 47.75 tons.
Cylinders: $16\frac{3}{8}$" x 24" (3).
Valve Gear: Bulleid chain driven (*Walschaerts. Piston valves).
Tractive Effort: 27 720 lbf.   **BR Power Classification:** 7P.

* Rebuilt at Eastleigh by Jarvis 1957–61 with the removal of the air-smoothed casing.
x Dual (air/vacuum) brakes.

BR	SR			
34007	21C107	WADEBRIDGE	Bodmin Steam Railway	1945
34010*	21C110	SIDMOUTH	Swanage Railway	1945 reb 1959
34016*	21C116	BODMIN	Mid Hants Railway	1945 reb 1958
34023	21C123	BLACKMORE VALE	Bluebell Railway	1946
34027* x	21C127	TAW VALLEY	Severn Valley Railway	1946 reb 1957
34028*	21C128	EDDYSTONE	Bluebell Railway	1946 reb 1958
34039*	21C139	BOSCASTLE	Great Central Railway	1946 reb 1959
34046* x	21C146	BRAUNTON	West Somerset Railway	1946 reb 1959
34051	21C151	WINSTON CHURCHILL	National Railway Museum	1946
34053*	21C153	SIR KEITH PARK	Hope Farm, Sellindge	1947 reb 1958
34058*	21C158	SIR FREDERICK PILE	Avon Valley Railway	1947 reb 1960
34059*	21C159	SIR ARCHIBALD SINCLAIR	Bluebell Railway	1947 reb 1960
34067 x	21C167	TANGMERE	Southall Depot, London	1947
34070	21C170	MANSTON	Swanage Railway	1947
34072		257 SQUADRON	Swanage Railway	1948
34073		249 SQUADRON	East Lancashire Railway	1948
34081		92 SQUADRON	North Norfolk Railway	1948
34092		CITY OF WELLS	Keighley & Worth Valley Railway	1949
34101*		HARTLAND	North Yorkshire Moors Railway	1950 reb 1960
34105		SWANAGE	Mid Hants Railway	1950

Notes:

34010 will be restored as "34109 SIR TRAFFORD LEIGH-MALLORY"
34023 was named BLACKMOOR VALE to 4/50.
34092 was named WELLS to 3/50.

CLASS MN   MERCHANT NAVY   4-6-2

Built: 1941–49. Bulleid SR design with air smoothed casing and similar features to 'WC' and 'BB'. All rebuilt 1956–59 by Jarvis to more conventional appearance. 30 built (35001–30). All locomotives were built and rebuilt at Eastleigh.
Boiler Pressure: 250 lbf/sq. in. superheated.   **Weight –Loco:** 97.9 tons.
Wheel Diameters: 3' 1", 6' 2", 3' 7".   **–Tender:** 47.8 tons.
Cylinders: 18" x 24" (3).   **Valve Gear:** Walschaerts. Piston valves.
Tractive Effort: 33 490 lbf.   **BR Power Classification:** 8P.

§ Sectioned.
x Dual (air/vacuum) brakes.

BR	SR			
35005	21C5	CANADIAN PACIFIC	Mid Hants Railway	1941 reb 1959
35006	21C6	PENINSULAR & ORIENTAL S.N. Co.	Gloucestershire-Warwickshire Rly	1941 reb 1959
35009	21C9	SHAW SAVILL	East Lancashire Railway	1942 reb 1957
35010	21C10	BLUE STAR	Colne Valley Railway	1942 reb 1957
35011	21C11	GENERAL STEAM NAVIGATION	HLPG, Binbrook Trading Estate	1944 reb 1959
35018	21C18	BRITISH INDIA LINE	GCE & SCS, Easton, Isle of Portland	1945 reb 1956
35022		HOLLAND AMERICA LINE	Southall Depot, London	1948 reb 1956
35025		BROCKLEBANK LINE	Great Central Railway	1948 reb 1956
35027		PORT LINE	Southall Depot, London	1948 reb 1957
35028 x		CLAN LINE	Stewarts Lane Depot, London	1948 reb 1959
35029 §		ELLERMAN LINES	National Railway Museum	1949 reb 1959

PLATFORM 5 MAIL ORDER

KEEP RIGHT UP TO DATE WITH THE CURRENT BRITISH ROLLING STOCK...

BRITISH RAILWAYS POCKET BOOKS 2008

The Platform 5 British Railways Pocket Books contain a full list of all vehicles in service with owner, operation, livery and depot information for every vehicle, plus pool codes for locomotives. Each book also includes an overview of Britain's railway network today and detailed lists of all train operating companies, depot and maintenance facilities, leasing companies and other useful information.

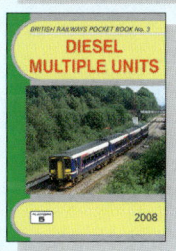

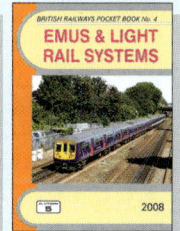

BRPB No. 1:	Locomotives	£4.25
BRPB No. 2:	Coaching Stock	£4.25
BRPB No. 3:	Diesel Multiple Units	£4.25
BRPB No. 4:	Electric Multiple Units	£4.25

A complete source of reference used throughout the railway industry.

HOW TO ORDER

Telephone your order and credit/debit card details to our 24-hour sales orderline:
0114 255 8000 or Fax: 0114 255 2471.
An answerphone is attached for calls made outside of normal UK office hours.
Or send your credit/debit card details, sterling cheque, money order or British Postal order payable to 'Platform 5 Publishing Ltd.' to:

Mail Order Department (PL), Platform 5 Publishing Ltd, 3 Wyvern House, Sark Road, SHEFFIELD, S2 4HG, ENGLAND

Please add postage & packing: 10% UK; 20% Europe; 30% Rest of World.
Please allow 28 days for delivery in the UK.

CLASS T3 — 4-4-0
Built: 1982–93. Adams LSWR design. 20 built
Boiler Pressure: 175 lbf/sq. in.
Wheel Diameters: 3' 7", 6' 7".
Cylinders: 19" x 26" (O).
Tractive Effort: 17 670 lbf.
Weight −Loco: 48.55 tons.
           −Tender: 33.2 tons.
Valve Gear: Stephenson. Slide valves.

SR   *LSWR*
E563–563   563   Locomotion: The NRM at Shildon   Nine Elms 380/1893

CLASS B1   "GLADSTONES"   0-4-2
Built: 1882–91. Stroudley LBSCR design. 49 built.
Boiler Pressure: 150 lbf/sq. in.
Wheel Diameters: 6' 6", 4' 6".
Cylinders: 18¼" x 26" (I).
Tractive Effort: 14 160 lbf.
Weight −Loco: 38.7 tons.
           −Tender: 29.35 tons.
Valve Gear: Stephenson. Slide valves.
Air brakes.

SR   *LBSCR*
B618   214–618   GLADSTONE   National Railway Museum   Brighton 1882

CANTERBURY & WHITSTABLE RAILWAY — 0-4-0
Built: 1830. Robert Stephenson & Company design.
Boiler Pressure: 40 lbf/sq. in.
Wheel Diameter: 4' 0".
Cylinders: 10½" x 18" (O).
Weight −Loco: 6.25 tons
           −Tender:
Tractive Effort: 2680 lbf.

INVICTA   Canterbury Heritage Centre   RS 24/1830

1.2.1 NEW BUILD PROJECTS
There is only one Southern new build project. This is the Brighton Atlantic. Technical details refer to the original locos.

CLASS H2 — 4-4-2
Original Class Built: 1911–12. 6 built (32421–32426). The new build loco uses the boiler from a GNR Atlantic which is of identical design to the LBSCR class.
Boiler Pressure: 170 lbf/sq. in.
Wheel Diameters: ?", 6'7½".
Cylinders: 21" x 26" (O).
Tractive Effort: 20 840 lbf.
Weight −Loco: 68.25 tons.
           −Tender: 39.25 tons.
Valve Gear: Stephenson. Piston valves.
BR Power Classification: 3P.

32424 BEACHY HEAD   Bluebell Railway   Under construction

LMS

1.3. LONDON MIDLAND & SCOTTISH RAILWAY & CONSTITUENT COMPANIES' STEAM LOCOMOTIVES

GENERAL

The LMS was formed in 1923 by the amalgamation of the Midland Railway, London & North Western Railway (LNWR), Caledonian Railway (CR), Glasgow & South Western Railway (GSWR) and Highland Railway (HR), plus a few smaller railways. Prior to this the North London Railway (NLR) and the Lancashire & Yorkshire Railway (L&Y) had been absorbed by the LNWR and the London, Tilbury & Southend Railway (LTSR) and been absorbed by the Midland Railway.

There is no new construction of LMS design locos under way, but a scheme has recently been announced to build an unrebuilt Patriot 4-6-0.

Numbering System

Originally number series were allocated to divisions as follows:

1– 4999	Midland Division (Midland and North Staffordshire Railway).
5000– 9999	Western Division "A" (LNWR).
10000–13999	Western Division "B" (L&Y).
14000–17999	Northern Division (Scottish Railways).

From 1934 onwards, all LMS standard locomotives and new builds were numbered in the range from 1–9999, and any locomotives which would have had their numbers duplicated had 20000 added to their original number.

At nationalisation 40000 was added to all LMS numbers except that locomotives which were renumbered in the 2xxxx series were further renumbered generally in the 58xxx series. In the following section locomotives are listed in order of BR number or in the position of the BR number they would have carried if they had lasted into BR days, except for very old locomotives which are listed at the end of the section.

Classification System

LMS locomotives did not generally have unique class designations but were referred to by their power classification which varied from 0 to 8 followed by the letters "P" for a passenger locomotive and "F" for a freight locomotive. Mixed traffic locomotives had no suffix letter. BR adopted the LMS system and used the description "MT" to denote these. The power classifications are generally shown in the class headings.

CLASS 4P   COMPOUND   4-4-0

Built: 1902–03. Johnson Midland design, rebuilt by Deeley 1914–19 to a similar design to the Deeley compounds which were built 1905–09. A further similar batch was built by the LMS in 1924–32. 240 built (41000–41199, 40900–40939).
Boiler Pressure: 200 lb/sq. in. superheated.   **Weight –Loco:** 61.7 tons.
Wheel Diameters: 3′ 6½″, 7′ 0″.   **–Tender:** 45.9 tons.
Cylinders: One high pressure. 19″ x 26″ (I).
                    Two low pressure. 21″ x 26″ (O).
Valve Gear: Stephenson. Slide valves on low pressure cylinders, piston valves on high pressure cylinder.
Tractive Effort: 23 205 lbf.

BR   LMS   MR
41000 1000   1000 (2361 pre-1907)   Severn Valley Railway (N)   Derby 1902 reb 1914

CLASS 2MT   2–6–2T

Built: 1946–52. Ivatt LMS design. 130 built (41200–41329).
Boiler Pressure: 200 lb/sq. in. superheated.   **Weight:** 65.2 (63.25*) tons.
Wheel Diameters: 3′ 0″, 5′ 0″, 3′ 0″.   **Cylinders:** 16½″ (16″*) x 24″ (O).
Valve Gear: Walschaerts. Piston valves.
Tractive Effort: 18 510 (17 410*) lbf.

41241*	Keighley & Worth Valley Railway	Crewe 1949
41298	Buckinghamshire Railway Centre	Crewe 1951
41312	Mid Hants Railway	Crewe 1952
41313	Isle of Wight Steam Railway	Crewe 1952

CLASS 1F   0-6-0T

Built: 1874–92. Johnson Midland design. Rebuilt with Belpaire boiler. 262 built.
Boiler Pressure: 150 lb/sq. in.   **Weight:** 45.45 tons.
Wheel Diameter: 4′ 6½″.   **Cylinders:** 17″ x 24″ (I).
Valve Gear: Stephenson. Slide valves.   **Tractive Effort:** 16 230 lbf.

BR   LMS   MR
41708 1708   1708   Barrow Hill Roundhouse   Derby 1880

LTS 79 CLASS (3P)   4-4-2T

Built: 1909. Whitelegg LTSR design. 4 built.
Boiler Pressure: 170 lb/sq. in.   **Weight:** 69.35 tons.
Wheel Diameters: 3′, 6″, 6′ 6″, 3′ 6″.   **Cylinders:** 19″ x 26″ (O).
Valve Gear: Stephenson. Slide valves.   **Tractive Effort:** 17 390 lbf.

BR   LMS   MR   LTSR
41966 2148   2177 80 THUNDERSLEY   Bressingham Steam Museum (N)   RS 3367/1909

CLASS 4MT   2-6-4T

Built: 1945–51. Fairburn modification of Stanier design (built 1936–43). This in turn was a development of a Fowler design built 1927–34. 383 built (Stanier & Fairburn). (42030–299/425–494/537–699).
Boiler Pressure: 200 lb/sq. in. superheated.   **Weight:** 85.25 tons.
Wheel Diameters: 3′, 3½″, 5′ 9″, 3′ 3½″.   **Cylinders:** 19¾″ x 26″ (O).
Valve Gear: Walschaerts. Piston valves.   **Tractive Effort:** 24 670 lbf.

42073	Lakeside & Haverthwaite Railway	Brighton 1950
42085	Lakeside & Haverthwaite Railway	Brighton 1951

CLASS 4MT 2-6-4T
Built: 1934. Stanier LMS 3-cylinder design for LTS line. 37 built (42500–36).
Boiler Pressure: 200 lb/sq. in. superheated.   **Weight:** 92.5 tons.
Wheel Diameters: 3', 3½", 5' 9", 3' 3½".   **Cylinders:** 16" x 26" (3).
Valve Gear: Walschaerts. Piston valves.   **Tractive Effort:** 24 600 lbf.

BR   LMS
42500 2500   National Railway Museum   Derby 1934

CLASS 5MT   "CRAB"   2-6-0
Built: 1926–32. Hughes LMS design. 245 built (42700–944).
Boiler Pressure: 180 lb/sq. in. superheated.   **Weight –Loco:** 66 tons.
Wheel Diameters: 3' 6½", 5' 6".   **–Tender:** 42.2 (41.5*) tons.
Cylinders: 21" x 26" (O).   **Valve Gear:** Walschaerts. Piston valves.
Tractive Effort: 26 580 lbf.

BR   LMS
42700   13000–2700   Locomotion: The NRM at Shildon   Horwich 1926
42765* 13065–2765   East Lancashire Railway   Crewe 5757/1927
42859   13159–2859   HLPG, Binbrook Trading Estate   Crewe 5981/1930

CLASS 5MT   2-6-0
Built: 1933–34. Stanier LMS design. 40 built (42945–84).
Boiler Pressure: 225 lb/sq. in. superheated.   **Weight –Loco:** 69.1 tons.
Wheel Diameters: 3' 3½", 5' 6".   **–Tender:** 42.2 tons.
Cylinders: 18" x 28" (O).   **Valve Gear:** Walschaerts. Piston valves.
Tractive Effort: 26 290 lbf.

BR   LMS
42968 13268–2968   Severn Valley Railway   Crewe 1934

CLASS 4MT   2-6-0
Built: 1947–52. Ivatt design. 162 built (43000–161).
Boiler Pressure: 225 lb/sq. in. superheated.   **Weight –Loco:** 59.1 tons.
Wheel Diameters: 3' 0", 5' 3".   **–Tender:** 40.3 tons.
Cylinders: 17½" x 26" (O).   **Valve Gear:** Walschaerts. Piston valves.
Tractive Effort: 24 170 lbf.

43106   Severn Valley Railway   Darlington 2148/1951

CLASS 4F   0-6-0
Built: 1911–41. Fowler Midland "Big Goods" design. Locomotives from 44027 onwards were LMS design with higher sided tenders. The preserved Midland locomotive has an LMS tender. 772 built (43835–44606).
Boiler Pressure: 175 lb/sq. in. superheated.   **Weight –Loco:** 48.75 tons.
Wheel Diameter: 5' 3".   **–Tender:** 41.2 tons
Cylinders: 20" x 26" (I).   **Valve Gear:** Stephenson. Piston valves.
Tractive Effort: 24 560 lbf.

BR   LMS   MR
43924 3924 3924   Keighley & Worth Valley Railway   Derby 1920
44027 4027   National Railway Museum   Derby 1924
44123 4123   Avon Valley Railway   Crewe 5658/1925
44422 4422   Churnet Valley Railway   Derby 1927

CLASS 5MT "BLACK 5" 4-6-0

Built: 1934–50. Stanier design. 842 built (45000–45499 then 44999–44658 in descending order).
Boiler Pressure: 225 lb/sq. in. superheated.   **Weight –Loco:** 72.1 (75.3*) tons.
Wheel Diameters: 3' 3½", 6' 0"   **–Tender:** 53.65 (53.8*) tons.
Cylinders: 18½" x 28" (O).   **Tractive Effort:** 25 450 lbf.
Valve Gear: Walschaerts. Piston valves. 44767 has outside Stephenson with piston valves.

x–Dual (air/vacuum) brakes.

BR	LMS			
44767	4767	"GEORGE STEPHENSON"	North Yorkshire Moors Railway	Crewe 1947
44806	4806	"MAGPIE"	Llangollen Railway	Derby 1944
44871	4871	"SOVEREIGN"	East Lancashire Railway	Crewe 1945
44901	4901		Barry Island Railway	Crewe 1945
44932	4932		Midland Railway-Butterley	Horwich 1945
45000	5000		National Railway Museum	Crewe 216/1935
45025	5025		Strathspey Railway	VF 4570/1934
45110	5110	"RAF BIGGIN HILL"	Severn Valley Railway	VF 4653/1935
45163	5163		Colne Valley Railway	AW 1204/1935
45212	5212	"ROY 'KORKY' GREEN"	North Yorkshire Moors Railway	AW 1253/1935
45231	5231		East Lancashire Railway	AW 1286/1935
45293	5293		Colne Valley Railway	AW 1348/1936
45305	5305		Great Central Railway	AW 1360/1937
45337	5337		East Lancashire Railway	AW 1392/1937
45379	5379		Mid Hants Railway	AW 1434/1937
45407 x	5407		East Lancashire Railway	AW 1462/1937
45428	5428	"ERIC TREACY"	North Yorkshire Moors Railway	AW 1483/1937
45491	5491		Midland Railway-Butterley	Derby 1943

Notes: 44767 is currently receiving attention at Ian Storey Engineering, Hepscott, near Morpeth.
45212 is on loan from the Keighley & Worth Valley Railway.

CLASS 6P (Formerly 5XP) JUBILEE 4-6-0

Built: 1934–36. Stanier taper boiler development of Patriot class. 191 built (45552–45742).
Boiler Pressure: 225 lb/sq. in. superheated.   **Weight –Loco:** 79.55 tons.
Wheel Diameters: 3' 3½", 6' 9".   **–Tender:** 53.65 tons.
Cylinders: 17" x 26" (3).   **Valve Gear:** Walschaerts. Piston valves.
Tractive Effort: 26 610 lbf.
* Fitted with double chimney.

BR	LMS			
45593	5593	KOLHAPUR	Barrow Hill Roundhouse	NBL 24151/1934
45596 *	5596	BAHAMAS	Keighley & Worth Valley Railway	NBL 24154/1935
45690	5690	LEANDER	East Lancashire Railway	Crewe 288/1936
45699	5699	GALATEA	West Coast Railway Company, Carnforth	Crewe 297/1936

CLASS 7P (Formerly 6P) ROYAL SCOT 4-6-0

Built: 1927–30. Fowler parallel design. All rebuilt 1943–55 with taper boilers and curved smoke deflectors. 71 built (46100–70).
Boiler Pressure: 250 lb/sq. in. superheated.   **Weight –Loco:** 83 tons.
Wheel Diameters: 3' 3½", 6' 9".   **–Tender:** 54.65 tons.
Cylinders: 18" x 26" (3).   **Valve Gear:** Walschaerts. Piston valves.
Tractive Effort: 33 150 lbf.

BR	LMS			
46100	6100	ROYAL SCOT	Southall Depot, London (N)	Derby 1930 reb Crewe 1947
46115	6115	SCOTS GUARDSMAN	West Coast Railway Company, Carnforth	NBL 23610/1927 reb Crewe 1950

Note: 6100 was built as 6152 THE KING'S DRAGOON GUARDSMAN. This loco swapped identities permanently with 6100 ROYAL SCOT in 1933 for a tour of the USA.

CLASS 8P (Formerly 7P) PRINCESS ROYAL 4-6-2
Built: 1933–35. Stanier design. 13 built (46200–12).
Boiler Pressure: 250 lb/sq. in. superheated. **Weight–Loco:** 105.5 tons.
Wheel Diameters: 3' 0", 6' 6", 3' 9".   **–Tender:** 54.65 tons.
Cylinders: 16¼" x 28" (4). **Valve Gear:** Walschaerts. Piston valves.
Tractive Effort: 40 290 lbf.

x–Dual (air/vacuum) brakes.

BR	LMS			
46201 x	6201	PRINCESS ELIZABETH	East Lancashire Railway	Crewe 107/1933
46203	6203	PRINCESS MARGARET ROSE	Midland Railway-Butterley	Crewe 253/1935

CLASS 8P (Formerly 7P) CORONATION 4-6-2
Built: 1937–48. Stanier design. 24 of this class were built streamlined but had the casing removed later. Certain locomotives were built with single chimneys, but all finished up with double chimneys. The tenders were fitted with steam driven coal-pushers. 38 built (46220–57).
Boiler Pressure: 250 lb/sq. in. superheated. **Weight –Loco:** 105.25 tons.
Wheel Diameters: 3' 0", 6' 9", 3' 9".   **–Tender:** 56.35 tons.
Cylinders: 16½" x 28" (4). **Valve Gear:** Walschaerts. Piston valves.
Tractive Effort: 40 000 lbf.

d–Formerly streamlined. Built with double chimney. Casing removed.
n–Never streamlined. Built with single chimney. Dual air/vacuum brakes.
s–Formerly streamlined. Built with single chimney. Casing removed, but new one being fitted.

BR	LMS			
46229 s	6229	DUCHESS OF HAMILTON	Tyseley locomotive Works (N)	Crewe 1938
46233 n	6233	DUCHESS OF SUTHERLAND	Midland Railway-Butterley	Crewe 1938
46235 d	6235	CITY OF BIRMINGHAM	Birmingham Museum of Science & Industry	Crewe 1939

CLASS 2MT 2-6-0
Built: 1946–53. Ivatt design. 128 built (46400–527).
Boiler Pressure: 200 lb/sq. in. superheated. **Weight –Loco:** 47.1 (48.45*) tons.
Wheel Diameters: 3' 0", 5' 0".   **–Tender:** 37.15 tons.
Cylinders: 16" (16½"*) x 24" (O). **Valve Gear:** Walschaerts. Piston valves.
Tractive Effort: 17 410 (18 510*) lbf.

46428	East Lancashire Railway	Crewe 1948
46441	Ribble Steam Railway	Crewe 1950
46443	Severn Valley Railway	Crewe 1950
46447	Buckinghamshire Railway Centre	Crewe 1950
46464	Caledonian Railway	Crewe 1950
46512* "E.V. COOPER ENGINEER"	Strathspey Railway	Swindon 1952
46521*	Great Central Railway	Swindon 1953

CLASS 3F "JINTY" 0-6-0T
Built: 1924–31. Fowler LMS development of his own Midland design. 422 built (47260–47681).
Boiler Pressure: 160 lb/sq. in. **Weight:** 49.5 tons.
Wheel Diameter: 4' 7". **Cylinders:** 18" x 26" (I).
Valve Gear: Stephenson. Slide valves. **Tractive Effort:** 20 830 lbf.

BR	LMS		
47279	7119–7279	Keighley & Worth Valley Railway	VF 3736/1924
47298	7138–7298	Llangollen Railway	HE 1463/1924
47324	16407–7324	East Lancashire Railway	NBL 23403/1926
47327	16410–7327	Midland Railway-Butterley	NBL 23406/1926
47357	16440–7357	Midland Railway-Butterley	NBL 23436/1926
47383	16466–7383	Severn Valley Railway	VF 3954/1926
47406	16489–7406	Great Central Railway	VF 3977/1926
47445	16528–7445	Midland Railway-Butterley	HE 1529/1927
47493	16576–7493	Spa Valley Railway	VF 4195/1928
47564	16647–7564	Midland Railway-Butterley	HE 1580/1928

Note: 47564 was latterly used as a stationary boiler numbered 2022.

CLASS 8F 2-8-0

Built: 1934–46. Stanier design. 331 built (8000–8225, 8301–8399, 8490–5). A further 521 were built to Ministry of Supply (208) Railway Executive Committee (245) and LNER (68) orders. Many of these operated on Britain's Railways with 228 being shipped overseas during the war. Post-war many were taken into LMS/BR stock including some returned from overseas
Boiler Pressure: 225 lb/sq. in. superheated. **Weight –Loco:** 72.1 tons.
Wheel Diameters: 3′ 3½″, ,4′ 8½″. **–Tender:** 53.65 tons.
Cylinders: 18½″ x 28″ (O). **Valve Gear:** Walschaerts. Piston valves.
Tractive Effort: 32 440 lbf. Turkish locos have air brakes.

*–Number allocated but never carried.
§–Became Persian Railways 41.109. WD number was 70307 in 1944 and 500 in 1952.

BR	LMS	WD	TCDD/IRR		
48151	8151			West Coast Railway Company, Carnforth	Crewe 1942
48173	8173			Avon Valley Railway	Crewe 1943
	8266*	340	45168	Izmit (plinthed), Turkey	NBL 24640/1940
	8274*	348	45160	Gloucestershire-Warwickshire Railway	NBL 24648/1940
	8279*	353	45165	Alaseher Station, Turkey	NBL 24653/1940
		522	45161	TCDD Steam Loco Museum, Camlik, Turkey	NBL 24670/1941
		547	1429	IRR, Baghdad, Iraq	NBL 24740/1942
48305	8305			The Railway Age, Crewe	Crewe 1943
48431	8431			Keighley & Worth Valley Railway	Swindon 1944
48518	8518			Barry Island Railway	Doncaster 1966/1944
48624	8624			Peak Railway	Ashford 1943
48773	8233	307§		Severn Valley Railway	NBL 24607/1940

LNWR CLASS G2 (7F) 0-8-0

Built: 1921–22. Beames development of 1912 Bowen-Cooke LNWR design. 60 built (49395–454). In addition many 1912 locos were rebuilt to similar condition.
Boiler Pressure: 175 lb/sq. in. superheated. **Weight –Loco:** 62 tons.
Wheel Diameter: 4′ 5½″. **–Tender:** 40.75 tons.
Cylinders: 20½″ x 24″ (I). **Valve Gear:** Joy. Piston valves.
Tractive Effort: 28 040 lbf.

BR	LMS	LNWR		
49395	9395	485	National Railway Museum	Crewe 5662/1921

L&Y CLASS 5 (2P) 2-4-2T

Built: 1889–1909. Aspinall L&Y design. 210 built.
Boiler Pressure: 180 lb/sq. in. **Weight:** 55.45 tons.
Wheel Diameters: 3′ 7¹/₈″, 5′ 7⁵/₈″, 3′ 7¹/₈″. **Cylinders:** 18″ x 26″ (I).
Valve Gear: Joy. Slide valves. **Tractive Effort:** 18 990 lbf.

BR	LMS	L&Y		
50621	10621	1008	National Railway Museum	Horwich 1/1889

L&Y CLASS 21 (0F) "PUG" 0-4-0ST

Built: 1891–1910. Aspinall L&Y design. 57 built.
Boiler Pressure: 160 lb/sq. in. **Weight:** 21.25 tons.
Wheel Diameter: 3′ 0¾″. **Cylinders:** 13″ x 18″ (O).
Valve Gear: Stephenson. Slide valves. **Tractive Effort:** 11 370 lbf.

BR	LMS	L&Y		
51218	11218	68	Keighley & Worth Valley Railway	Horwich 811/1901
	11243	19	Ribble Steam Railway	Horwich 1097/1910

L&Y CLASS 23 (2F)   0-6-0ST
Built: 1891–1900. Aspinall rebuild of Barton Wright L&Y 0-6-0. 230 built.
Boiler Pressure: 140 lb/sq. in.   **Weight:** 43.85 tons
Wheel Diameter: 4' 5$^{7}/_{8}$".   **Cylinders:** 17½" x 26" (I).
Valve Gear: Joy. Slide valves.   **Tractive Effort:** 17 590 lbf.

BR	LMS	L&Y		
–	11456	752	Keighley & Worth Valley Railway	BP 1989/1881 reb. Hor. 1896

L&Y CLASS 25 (2F)   0-6-0
Built: 1876–87. Barton Wright L&Y design. 280 built.
Boiler Pressure: 140 lb/sq. in.   **Weight –Loco:** 39.05 tons.
Wheel Diameter: 4' 5$^{7}/_{8}$".   **–Tender:** 28.5 tons.
Cylinders: 17½" x 26" (I).   **Valve Gear:** Joy. Slide valves.
Tractive Effort: 17 590 lbf.

BR	LMS	L&Y		
52044	12044	957	Keighley & Worth Valley Railway	BP 2840/1887

L&Y CLASS 27 (3F)   0-6-0
Built: 1889–1917. Aspinall L&Y design. 448 built.
Boiler Pressure: 180 lb/sq. in.   **Weight –Loco:** 44.3 tons.
Wheel Diameter: 5" 0$^{7}/_{8}$".   **–Tender:** 26.1 tons.
Cylinders: 18" x 26" (I).   **Valve Gear:** Joy. Slide valves.
Tractive Effort: 21 170 lbf.

BR	LMS	L&Y		
52322	12322	1300	Ribble Steam Railway	Horwich 420/1896

CLASS 7F   2-8-0
Built: 1914–25. Fowler design for Somerset & Dorset Joint Railway (Midland and LSWR jointly owned). 11 built (53800–10).
Boiler Pressure: 190 lb/sq. in. superheated.   **Weight –Loco:** 64.75 tons.
Wheel Diameters: 3' 3½", 4' 7½".   **–Tender:** 26.1 tons.
Cylinders: 21" x 28" (O).   **Valve Gear:** Walschaerts. Piston valves.
Tractive Effort: 35 950 lbf.

BR	LMS	S&DJR		
53808	9678–13808	88	West Somerset Railway	RS 3894/1925
53809	9679–13809	89	Midland Railway-Butterley	RS 3895/1925

Note: 53809 often carries the name "BEAUMONT".

CALEDONIAN RAILWAY (1P)   4–2–2
Built: 1886. Drummond design. 1 built.
Boiler Pressure: 160 lb/sq. in.   **Weight –Loco:** 41.35 tons.
Wheel Diameters: 3' 6", 7' 0", 4' 6".   **–Tender:** 35.4 tons.
Cylinders: 18" x 26" (I).   **Valve Gear:** Stephenson. Slide valves.
Tractive Effort: 13 640 lbf.

BR	LMS	CR		
–	14010	123	Glasgow Museum of Transport	N 3553/1886

CALEDONIAN RAILWAY 439 CLASS (2P)   0-4-4T
Built: 1900–14. McIntosh design. 68 built.
Boiler Pressure: 160 lb/sq. in.   **Weight:** 53.95 tons.
Wheel Diameters: 5' 9", 3' 2".   **Cylinders:** 18" x 26" (I).
Valve Gear: Stephenson. Slide valves.   **Tractive Effort:** 16 600 lbf.
Dual brakes.

BR	LMS	CR		
55189	15189	419	Bo'ness & Kinneil Railway	St. Rollox 1908

GSWR 322 CLASS (3F)   0-6-0T
Built: 1917. Drummond design. 3 built.
Boiler Pressure: 160 lb/sq. in.
Wheel Diameter: 4' 2".
Valve Gear: Walschaerts. Piston valves.
Weight: 40 tons.
Cylinders: 17" x 22" (O).
Tractive Effort: 17 290 lbf.

BR	LMS	GSWR		
–	16379	9	Glasgow Museum of Transport	NBL 21521/1917

CALEDONIAN RAILWAY 812 CLASS (3F)   0-6-0
Built: 1899–1900. McIntosh design. 96 built (57550–645).
Boiler Pressure: 160 lb/sq. in.
Wheel Diameter: 5' 0".
Cylinders: 18½" x 26" (I).
Tractive Effort: 20 170 lbf.
Weight –Loco: 45.7 tons.
         –Tender: 37.9 tons.
Valve Gear: Stephenson. Slide valves.
Air brakes.

BR	LMS	CR		
57566	17566	828	Strathspey Railway	St. Rollox 1899

HIGHLAND RAILWAY (4F)   "JONES GOODS"   4-6-0
Built: 1894. Jones design. 15 built.
Boiler Pressure: 175 lb/sq. in.
Wheel Diameters: 3' 3", 5' 3".
Cylinders: 20" x 26" (O).
Tractive Effort: 24 560 lbf.
Weight –Loco: 56 tons.
         –Tender: 38.35 tons.
Valve Gear: Stephenson. Slide valves.

BR	LMS	HR		
–	17916	103	Glasgow Museum of Transport	SS 4022/1894

NORTH LONDON RAILWAY 75 CLASS (2F)   0-6-0T
Built: 1881–1905. Park design. 15 built.
Boiler Pressure: 160 lb/sq. in.
Wheel Diameter: 4' 4".
Valve Gear: Stephenson. Slide valves.
Weight: 45.55 tons.
Cylinders: 17" x 24" (O).
Tractive Effort: 18 140 lbf.

BR	LMS	LNWR	NLR		
58850	7505–27505	2650	116	Bluebell Railway	Bow 181/1881

LNWR COAL TANK (1F)   0-6-2T
Built: 1881–1896. Webb design. 300 built.
Boiler Pressure: 150 lb/sq. in.
Wheel Diameters: 4' 5½", 3' 9".
Valve Gear: Stephenson. Slide valves.
Weight: 43.75 tons
Cylinders: 17" x 24" (I).
Tractive Effort: 16 530 lbf.

BR	LMS	LNWR		
58926	7799	1054	Keighley & Worth Valley Railway	Crewe 2979/1888

MIDLAND 156 CLASS (1P)   2-4-0
Built: 1866–68. Kirtley design. 23 built.
Boiler Pressure: 140 lb/sq. in.
Wheel Diameters: 4' 3", 6' 3".
Cylinders: 18" x 24" (I).
Tractive Effort: 12 340 lbf
Weight –Loco: 41.25 tons.
         –Tender: 34.85 tons.
Valve Gear: Stephenson. Slide valves.

BR	LMS	MR		
–	2-20002	158–158A	Locomotion: The NRM at Shildon	Derby 1866

MIDLAND 115 CLASS   "SPINNER"   4–2–2
Built: 1887–1900. Johnson design. 95 built.
Boiler Pressure: 170 lb/sq. in.
Wheel Diameters: 3′ 10″, 7′ 9½″, 4′ 4½″
Cylinders: 19″ x 26″ (I).
Tractive Effort: 15 280 lbf.
Weight –Loco: 43.95 tons.
        –Tender: 21.55 tons.
Valve Gear: Stephenson. Slide valves.

LMS	MR		
673	118–673	National Railway Museum	Derby 1897

NORTH STAFFS RAILWAY New L CLASS   0-6-2T
Built: 1903–23. Hookham design. 34 built.
Boiler Pressure: 175 lb/sq. in.
Wheel Diameters: 5′ 0″, 4′ 0″.
Valve Gear: Stephenson. Slide valves.
Weight: 64.95 tons.
Cylinders: 18½″ x 26″ (I).
Tractive Effort: 22 060 lbf

LMS	NSR		
2271	2	Locomotion: The NRM at Shilldon	Stoke 1923

Note: Although built in 1923, this loco carried an NSR number, since the NSR was not taken over by the LMS until late 1923.

LNWR   PRECEDENT   2-4-0
Built: 1874–82. Webb design. 166 built.
Boiler Pressure: 150 lb/sq. in.
Wheel Diameters: 3′ 9″, 6′ 9″.
Cylinders: 17″ x 24″ (I).
Tractive Effort: 10 920 lbf
Weight –Loco: 35.6 tons.
        –Tender: 25 tons.
Valve Gear: Allan.

LMS	LNWR			
5031	790	HARDWICKE	National Railway Museum	Crewe 3286/1892

LNWR   2-2-2
Built: 1847. Trevithick design rebuilt by Ramsbottom in 1858.
Boiler Pressure: 140 lb/sq. in.
Wheel Diameters: 3′ 6″, 8′ 6″, 3′ 6″.
Cylinders: 17¼″ x 24″ (O).
Tractive Effort: 8330 lbf
Weight –Loco: 29.9 tons.
        –Tender: 25 tons.
Valve Gear: Stephenson. Slide valves.

LNWR			
173–3020	CORNWALL	Locomotion: The NRM at Shilldon	Crewe 35/1847

LNWR   0-4-0ST
Built: 1865. Ramsbottom design.
Boiler Pressure: 120 lb/sq. in.
Wheel Diameter: 4′ 0″.
Tractive Effort: 8330 lbf.
Weight: 22.75 tons.
Cylinders: 14″ x 20″ (I).

LNWR			
1439–1985–3042		Locomotion: The NRM at Shilldon	Crewe 842/1865

GRAND JUNCTION RAILWAY   2-2-2
Built: 1845. Trevithick design.
Boiler Pressure: 120 lb/sq. in.
Wheel Diameters: 3′ 6″, 6′ 0″, 3′ 6″.
Cylinders: 15″ x 20″ (O).
Tractive Effort: 6375 lbf.
Weight –Loco: 20.4 tons.
        –Tender: 16.4 tons.
Valve Gear: Allan.

LNWR	GJR		
49	49–1868 COLUMBINE	Science Museum, London (N)	Crewe 25/1845

FURNESS RAILWAY 0-4-0
Built: 1846.
Boiler Pressure: 110 lb/sq. in.
Wheel Diameter: 4' 9".
Cylinders: 14" x 24" (I).
Tractive Effort: 7720 lbf.
Weight –Loco: 20 tons.
    –Tender: 13 tons.
Valve Gear: Stephenson. Slide valves.

| 3 | COPPERNOB | National Railway Museum | BCK 1846 |

FURNESS RAILWAY 0-4-0
Built: 1863 as 0-4-0. Sold in 1870 to Barrow Steelworks and numbered 7. Rebuilt 1915 as 0-4-0ST. Restored to original condition 1999.
Boiler Pressure: 120 lb/sq. in.
Wheel Diameter: 4'10"
Tractive Effort: 10140 lbf.
Weight:
Cylinders: 15½" x 24"(I)

| 20 | | Lakeside & Haverthwaite Railway | SS 1435/1863 |

FURNESS RAILWAY 0-4-0ST
Built: 1865 as 0-4-0. Sold in 1873 to Barrow Steelworks and numbered 17. Rebuilt 1921 as 0-4-0ST.
Boiler Pressure: 120 lb/sq. in.
Wheel Diameter: 4'3"
Tractive Effort: 11532 lbf.
Weight:
Cylinders: 15½" x 24"(I)

FR   Present
| 25 | 6 | West Coast Railway Company, Carnforth | SS 1585/1865 |

LIVERPOOL & MANCHESTER RAILWAY 0–2–2
Built: 1829 for the Rainhill trials.
Boiler Pressure: 50 lb/sq. in.
Wheel Diameters: 4' 8½", 2' 6".
Cylinders: 8" x 17" (O).
Weight –Loco: 4.25 tons.
    –Tender: 5.2 tons.
Tractive Effort: 820 lbf.

| ROCKET | Science Museum, London (N) | RS 1/1829 |

LIVERPOOL & MANCHESTER RAILWAY 0-4-0
Built: 1829 for the Rainhill trials.
Boiler Pressure: 50 lb/sq. in.
Wheel Diameters: 4' 6".
Cylinders: 7" x 18" (O).
Weight –Loco: 4.25 tons.
    –Tender: 5.2 tons.
Tractive Effort: 690 lbf.

| SANS PAREIL | Locomotion: The NRM at Shildon | Hack 1829 |

LIVERPOOL & MANCHESTER RAILWAY 0–4–2
Built: 1838–9. Four built. The survivor was the star of the film "The Titfield Thunderbolt".
Boiler Pressure: 50 lb/sq. in.
Wheel Diameters: 5' 0", 3' 3".
Cylinders: 14" x 24" (I).
Weight –Loco: 14.45 tons.
    –Tender:
Tractive Effort: 3330 lbf.

L&MR   LNWR
| 57 | 116 | LION | Liverpool Museum Store, Bootle | TKL 1838 |

MERSEY RAILWAY 0–6–4T
Built: 1885. Withdrawn 1903 on electrification. No. 5 was sold to Coppice colliery in Derbyshire.
Boiler Pressure: 150 lb/sq. in.
Wheel Diameters: 4' 7", 3' 0".
Valve Gear: Stephenson. Slide valves.
Weight: 67.85 tons.
Cylinders: 21" x 26" (I).
Tractive Effort: 26 600 lbf.

| 1 | THE MAJOR | Rail Transport Museum, Thirlmere, NSW, Australia | BP 2601/1885 |
| 5 | CECIL RAIKES | Liverpool Museum Store, Bootle | BP 2605/1885 |

LNER

1.4. LONDON & NORTH EASTERN RAILWAY AND CONSTITUENT COMPANIES' STEAM LOCOMOTIVES

GENERAL

The LNER was formed in 1923 by the amalgamation of the Great Northern Railway (GNR). North Eastern Railway (NER), Great Eastern Railway (GER), Great Central Railway (GCR), North British Railway (NBR) and Great North of Scotland Railway (GNSR). Prior to this the Hull & Barnsley Railway (H & B) had been absorbed by the NER in 1922.

Locomotive Numbering System

Initially pre grouping locomotive numbers were retained, but in September 1923 suffix letters started to be applied depending upon the works which repaired the locomotives. In 1924 locomotives were renumbered in blocks as follows: NER locomotives remained unaltered, GNR locomotives had 3000 added, GCR 5000, GNSR 6800, GER 7000 and NBR 9000. New locomotives filled in gaps between existing numbers. By 1943 the numbering of new locomotives had become so haphazard that it was decided to completely renumber locomotives so that locomotives of a particular class were all contained in the same block of numbers. On nationalisation in 1948, 60000 was added to LNER numbers.

Classification System

The LNER gave each class a unique code consisting of a letter denoting the wheel arrangement and a number denoting the individual class within the wheel arrangement. Route availability (RA) was denoted by a number, the higher the number the more restricted the route availability.

CLASS A4   4-6-2

Built: 1935–38. Gresley streamlined design. "MALLARD" attained the world speed record for a steam locomotive of 126.4 mph in 1938 and is still unbeaten. 35 built (2509–12, 4462–69/82–4500/4900–03).
Boiler Pressure: 250 lb/sq. in. superheated.   **Weight –Loco:** 102.95 tons.
Wheel Diameters: 3' 2", 6' 8", 3' 8".   **–Tender:** 64.15 tons.
Cylinders: 18½" x 26" (3).   **Tractive Effort:** 35 450 lbf.
Valve Gear: Walschaerts with derived motion for inside cylinder. Piston valves.
BR Power Classification: 8P.   **RA:** 9.
x–Dual (air/vacuum) brakes.

BR	LNER			
60007	4498–7	SIR NIGEL GRESLEY	North Yorkshire Moors Railway	Doncaster 1863/1937
60008	4496–8	DWIGHT D. EISENHOWER	National RR Museum, USA	Doncaster 1861/1937
60009x	4488–9	UNION OF SOUTH AFRICA	Thornton Depot, Fife	Doncaster 1853/1937
60010	4489–10	DOMINION OF CANADA	Canadian RR Historical Museum	Doncaster 1854/1937
60019	4464–19	BITTERN	Mid Hants Railway	Doncaster 1866/1937
60022	4468–22	MALLARD	National Railway Museum	Doncaster 1870/1937

CLASS A3   4-6-2

Built: 1922–35. Gresley design. Built as Class A1 (later reclassified A10 after the Peppercorn A1s were being designed), but rebuilt to A3 in 1947. 79 built. (60035–113). Rebuilt 1959 with Kylchap blastpipe, double chimney and German-style smoke deflectors. Restored to original state on preservation in 1963, but restored to rebuilt state in 1993. The loco is numbered 4472, a number it never carried as an A3, since it was renumbered 103 in 1946.
Boiler Pressure: 220 lb/sq. in. superheated.   **Weight –Loco:** 96.25 tons.
Wheel Diameters: 3' 2", 6' 8", 3' 8".   **–Tender:** 62.4 tons.
Cylinders: 19" x 26" (3).   **Tractive Effort:** 32 910 lbf.
Valve Gear: Walschaerts with derived motion for inside cylinder. Piston valves.
BR Power Classification: 7P.   **RA:** 9.
Dual (air/vacuum) brakes.

BR	LNER			
60103	1472–4472–502–103	FLYING SCOTSMAN	National Railway Museum	Doncaster 1564/1923

CLASS A2   4-6-2

Built: 1944–48. Peppercorn development of Thompson design. 40 built (60500–39).
Boiler Pressure: 250 lb/sq. in. superheated.   **Weight –Loco:** 101 tons.
Wheel Diameters: 3' 2", 6' 2", 3' 8".   **–Tender:** 60.35 tons.
Cylinders: 19" x 26" (3).   **Valve Gear:** Walschaerts. Piston valves.
Tractive Effort: 40 430 lbf.   **BR Power Classification:** 8P.
RA: 9.

60532		BLUE PETER	Barrow Hill Roundhouse	Doncaster 2023/1948

CLASS V2   2-6-2

Built: 1936–44. Gresley design for express passenger and freight. 184 built (60800–983).
Boiler Pressure: 220 lb/sq. in. superheated.   **Weight –Loco:** 93.1 tons.
Wheel Diameters: 3' 2", 6' 2", 3' 8".   **–Tender:** 52 tons.
Cylinders: 18½" x 26" (3).   **Tractive Effort:** 33 730 lbf.
Valve Gear: Walschaerts with derived motion for inside cylinder. Piston valves.
BR Power Classification: 6MT.   **RA:** 6.

BR	LNER			
60800	4771–800	GREEN ARROW	National Railway Museum	Doncaster 1837/1936

CLASS B1   4-6-0
Built: 1942–51. Thompson design. 410 built (61000–409).
Boiler Pressure: 225 lb/sq. in. superheated.   **Weight –Loco:** 71.15 tons.
Wheel Diameters: 3' 2", 6' 2".   **–Tender:** 52 tons.
Cylinders: 20" x 26" (O).   **Valve Gear:** Walschaerts. Piston valves.
Tractive Effort: 26 880 lbf   **BR Power Classification:** 5MT.
RA: 5.

BR	LNER	Present		
61264	1264		Barrow Hill Roundhouse	NBL 26165/1947
61306		1306 "MAYFLOWER"	Nene Valley Railway	NBL 26207/1948

Notes: The original MAYFLOWER was 61379. 61264 also carried Departmental 29.

CLASS B12   4-6-0
Built: 1911–28. Holden GER design. 80 built (8500–5/7–80). GER class S69.
Boiler Pressure: 180 lb/sq. in. superheated.   **Weight –Loco:** 69.5 tons.
Wheel Diameters: 3' 3", 6' 6".   **–Tender:** 39.3 tons.
Cylinders: 20" x 28" (I).   **Valve Gear:** Stephenson. Slide valves.
Tractive Effort: 21 970 lbf.   **BR Power Classification:** 4P.
RA: 5.   Dual brakes.

BR	LNER	GER		
61572	8572	1572	North Norfolk Railway	BP 6488/1928

CLASS K4   2-6-0
Built: 1937–38. Gresley design for West Highland line. 6 built (61993–8).
Boiler Pressure: 200 lb/sq. in. superheated.   **Weight –Loco:** 68.4 tons.
Wheel Diameters: 3' 2", 5' 2".   **–Tender:** 44.2 tons.
Cylinders: 18½" x 26" (3).   **Tractive Effort:** 36 600 lbf.
Valve Gear: Walschaerts with derived motion for inside cylinder. Piston valves.
BR Power Classification: 6MTP.   **RA:** 6.

BR	LNER			
61994	3442–1994	THE GREAT MARQUESS	Thornton Depot, Fife	Darlington 1761/1938

CLASS K1   2-6-0
Built: 1949–50. Peppercorn design. 70 built (62001–70).
Boiler Pressure: 225 lb/sq. in. superheated.   **Weight –Loco:** 66 tons.
Wheel Diameters: 3' 2", 5' 2".   **–Tender:** 52.2 tons.
Cylinders: 20" x 26" (O).   **Valve Gear:** Walschaerts. Piston valves.
Tractive Effort: 32 080 lbf.   **BR Power Classification:** 6MT.
RA: 6.

BR	Present		
62005	2005	North Yorkshire Moors Railway	NBL 26609/1949

CLASS D40   4-4-0
Built: 1889–1921. Pickersgill GNSR class F. 20 built.
Boiler Pressure: 165 lb/sq. in. superheated.   **Weight –Loco:** 48.65 tons.
Wheel Diameters: 3' 9½", 6' 1".   **–Tender:** 37.4 tons
Cylinders: 18" x 26" (I).   **Valve Gear:** Stephenson. Slide valves.
Tractive Effort: 16 180 lbf.   **BR Power Classification:** 1P.
RA: 4.

BR	LNER	GNSR		
62277	6849–2277	49	GORDON HIGHLANDER` Glasgow Museum of Transport	NBL 22563/1920

CLASS D34   GLEN   4-4-0

Built: 1913–20. Reid NBR class K. 32 built.
Boiler Pressure: 165 lb/sq. in. superheated.
Wheel Diameters: 3' 6", 6' 0".
Cylinders: 20" x 26" (I).
Tractive Effort: 22 100 lbf.
RA: 6.
Weight –Loco: 57.2 tons.
          –Tender: 46.65 tons.
Valve Gear: Stephenson. Piston valves.
BR Power Classification: 3P.

BR	LNER	NBR			
62469	9256–2469	256	GLEN DOUGLAS	Bo'ness & Kinneil Railway	Cowlairs 1913

CLASS D11   IMPROVED DIRECTOR   4-4-0

Built: 1919–22. Robinson GCR class 11F. 11 built (62660–70). 24 similar locomotives were built by the LNER.
Boiler Pressure: 180 lb/sq. in. superheated.
Wheel Diameters: 3' 6", 6' 9".
Cylinders: 20" x 26" (I).
Tractive Effort: 19 640 lbf.
RA: 6.
Weight –Loco: 61.15 tons.
          –Tender: 48.3 tons.
Valve Gear: Stephenson. Piston valves.
BR Power Classification: 3P.

BR	LNER	GCR			
62660	5506–2660	506	BUTLER HENDERSON	Barrow Hill Roundhouse	Gorton 1919

CLASS D49   4-4-0

Built: 1927. Gresley design. 76 built (62700–75).
Boiler Pressure: 180 lb/sq. in. superheated.
Wheel Diameters: 3' 1¼", 6' 8".
Cylinders: 17" x 26" (3).
Valve Gear: Walschaerts with derived motion for inside cylinder. Piston valves.
BR Power Classification: 4P.
Weight –Loco: 66 tons.
          –Tender: 52 tons.
Tractive Effort: 21 560 lbf.
RA: 8.

BR	LNER			
62712	246–2712	MORAYSHIRE	Bo'ness & Kinneil Railway	Darlington 1391/1928

CLASS E4   2-4-0

Built: 1891–1902. Holden GER class T26. 100 built.
Boiler Pressure: 160 lb/sq. in.
Wheel Diameters: 4' 0", 5' 8".
Cylinders: 17½" x 24" (I).
Tractive Effort: 14 700 lbf
RA: 2.
Weight –Loco: 40.3 tons.
          –Tender: 30.65 tons.
Valve Gear: Stephenson. Slide valves.
BR Power Classification: 1MT.
Air brakes.

BR	LNER	GER		
62785	7490–7802–2785	490	Bressingham Steam Museum (N)	Stratford 836/1894

CLASS C1   4-4-2

Built: 1902–10. Ivatt GNR class C1. 94 built.
Boiler Pressure: 170 lb/sq. in. superheated.
Wheel Diameters: 3' 8", 6' 8", 3' 8".
Cylinders: 20" x 24" (O).
Tractive Effort: 17 340 lbf.
RA: 7.
Weight –Loco: 69.6 tons.
          –Tender: 43.1 tons.
Valve Gear: Stephenson. Slide valves.
BR Power Classification: 2P.

BR	LNER	GNR		
–	3251–2800	251	Bressingham Steam Museum	Doncaster 991/1902

CLASS Q6   0-8-0
Built: 1913–21. Raven NER class T2. 120 built (63340–459).
Boiler Pressure: 180 lb/sq. in. superheated.   **Weight –Loco:** 65.9 tons.
Wheel Diameter: 4' 7¼".   **–Tender:** 44.1 tons.
Cylinders: 20" x 26" (O).   **Valve Gear:** Stephenson. Piston valves.
Tractive Effort: 28 800 lbf.   **BR Power Classification:** 6 F.
RA: 6.

BR	LNER	NER		
63395	2238–3395	2238	North Yorkshire Moors Railway	Darlington 1918

CLASS Q7   0-8-0
Built: 1919–24. Raven NER class T3. 15 built (63460–74).
Boiler Pressure: 180 lb/sq. in. superheated.   **Weight –Loco:** 71.6 tons.
Wheel Diameter: 4' 7¼".   **–Tender:** 44.1 tons.
Cylinders: 18½" x 26" (3).   **Valve Gear:** Stephenson. Piston valves..
Tractive Effort: 36 960 lbf.   **BR Power Classification:** 8F.
RA: 7.

BR	LNER	NER		
63460	901–3460	901	Locomotion: The NRM at Shildon	Darlington 1919

CLASS O4   2-8-0
Built: 1911–20. Robinson GCR class 8K 70 built. A further 521 were built, being ordered by the railway operating department (ROD). These saw service on British and overseas railways during and after the first world war. Some subsequently passed to British railway administrations whilst others were sold abroad.
Boiler Pressure: 180 lb/sq. in. superheated.   **Weight –Loco:** 73.2 tons.
Wheel Diameters: 3' 6", 4' 8".   **–Tender:** 48.3 tons.
Cylinders: 21" x 26" (O).   **Valve Gear:** Stephenson. Piston valves.
Tractive Effort: 31 330 lbf.   **BR Power Classification:** 7F.
RA: 6.

BR	LNER	GCR		
63601	5102–3509–3601	102	Great Central Railway (N)	Gorton 1911
ROD				
1984	Dorrigo, Northern New South Wales, Australia			NBL 22042/1918
2003	Dorrigo, Northern New South Wales, Australia			Gorton 1918
2004	Richmond Vale Steam Centre, Kurri-Kurri, NSW, Australia			Gorton 1918

CLASS J21   0-6-0
Built: 1886–95. Worsdell NER class C. 201 built.
Boiler Pressure: 160 lb/sq. in. superheated.   **Weight –Loco:** 43.75 tons.
Wheel Diameter: 5' 1¼".   **–Tender:** 36.95 tons.
Cylinders: 19" x 24" (I).   **Valve Gear:** Stephenson. Piston valves.
Tractive Effort: 19 240 lbf.   **BR Power Classification:** 2F.
RA: 3.

BR	LNER	NER		
65033	876–5033	876	North Norfolk Railway	Gateshead 1889

CLASS J36   0-6-0
Built: 1889–1900. Holmes NBR class C. 168 built.
Boiler Pressure: 165 lb/sq. in.   **Weight –Loco:** 41.95 tons.
Wheel Diameter: 5' 0".   **–Tender:** 33.5 tons.
Cylinders: 18" x 26" (I).   **Valve Gear:** Stephenson. Slide valves.
Tractive Effort: 20 240 lbf.   **BR Power Classification:** 2F.
RA: 3.

BR	LNER	NBR			
65243	9673–5243	673	MAUDE	Bo'ness & Kinneil Railway	N 4392/1891

CLASS J15   0-6-0

Built: 1883–1913. Worsdell GER class Y14. 189 built.
Boiler Pressure: 160 lb/sq. in.
Wheel Diameter: 4' 11".
Cylinders: 17½" x 24" (I).
Tractive Effort: 16 940 lbf.
RA: 1.
Weight –Loco: 37.1 tons.
        –Tender: 30.65 tons.
Valve Gear: Stephenson. Slide valves.
BR Power Classification: 2F.
Dual brakes.

BR	LNER	GER		
65462	7564–5462	564	North Norfolk Railway	Stratford 1912

CLASS J17   0-6-0

Built: 1900–11. Holden GER class G58. 90 built (65500–89).
Boiler Pressure: 180 lb/sq. in. superheated.
Wheel Diameter: 4' 11".
Cylinders: 19" x 26" (I).
Tractive Effort: 24 340 lbf.
RA: 4.
Weight –Loco: 45.4 tons.
        –Tender: 38.25 tons.
Valve Gear: Stephenson. Slide valves.
BR Power Classification: 4F.
Air brakes.

BR	LNER	GER		
65567	8217–5567	1217	National Railway Museum	Stratford 1905

CLASS J27   0-6-0

Built: 1906–23. Worsdell NER class P3. 115 built.
Boiler Pressure: 180 lb/sq. in. superheated.
Wheel Diameter: 4' 7¼".
Cylinders: 18½" x 26" (I).
Tractive Effort: 24 640 lbf.
RA: 5.
Weight –Loco: 47 tons.
        –Tender: 37.6 tons.
Valve Gear: Stephenson. Piston valves.
BR Power Classification: 4F.

BR	LNER		
65894	2392–5894	NELPG, Hopetown, Darlington	Darlington 1923

CLASS J94. 68077/8 (LNER 8077/8) – see War Department Steam Locomotives.

CLASS Y5   0-4-0ST

Built: 1874–1903. Neilson & Company design for GER (Class 209). 8 built. Survivor sold in 1917.
Boiler Pressure: 140 lb/sq. in.
Wheel Diameter: 3' 7".
Valve Gear: Stephenson. Slide valves.
RA: 1.
Weight: 21.2 tons.
Cylinders: 12" x 20" (O).
Tractive Effort: 7970 lbf.

GER		
229	North Woolwich Station Museum	N 2119/1876

CLASS Y7   0-4-0T

Built: 1888–1923. Worsdell NER class H. 24 built.
Boiler Pressure: 160 lb/sq. in.
Wheel Diameter: 4' 0".
Cylinders: 14" x 20" (I).
BR Power Classification: 0 F.
* No train brakes.
Weight: 22.7 tons.
Tractive Effort: 11 140 lbf.
Valve Gear: Joy. Slide valves.
RA: 1.

BR	LNER	NER		
68088*	985–8088		North Norfolk Railway	Darlington 1205/1923
	1310	1310	Middleton Railway	Gateshead 38/1891

CLASS Y9   0-4-0ST

Built: 1882–99. Drummond NBR class G. 35 built.
Boiler Pressure: 130 lb/sq. in.
Wheel Diameter: 3' 8".
Valve Gear: Stephenson. Slide valves.
BR Power Classification: 0F.
Weight: 27.8 tons.
Cylinders: 14" x 20" (I).
Tractive Effort: 9840 lbf.
RA: 2.

BR	LNER	NBR		
68095	9042–8095	42	Bo'ness & Kinneil Railway	Cowlairs 1887

▲ GWR 4900 Class 4-6-0 No. 4936 KINLET HALL south of Crowcombe Heathfield on the West Somerset Railway with a train for Minehead on 18 March 2005. **Paul Chancellor**

▼ GWR 6000 Class 4-6-0 No. 6024 KING EDWARD I, based at Didcot Railway Centre with the 15.05 Bishops Lydeard–Minehead at Leigh Woods on the West Somerset Railway on 20 March 2005. **Charles Woodland**

▲ GWR Dukedog 4-4-0 No. 9017 EARL OF BERKELEY on Freshfield Bank, Bluebell Railway on 4 November 2004. **Alan Barnes**

▼ A new locomotive! The West Somerset Railway's 4300 Class 2-6-0 No. 9351, converted from a 5101 Class 2-6-2T is seen at Arley on the Severn Valley Railway on 24 September 2005 with a demonstration freight service for Bewdley. **Paul Chancellor**

▲ Rebuilt West Country Pacifid No. 34028 EDDYSTONE works past New Barn Farm with the 16.20 Swanage–Norden, Swanage Railway Service (carrying a headboard commemorating the end of Southern Steam, 9/7/67) on 9 July 2005. **Rail Photoprints**

▼ An unrebuilt one, No. 34067 TANGMERE works the 14.22 Penzance–Exeter St. Davids special at Cockwood Harbour near Starcross on 29 August 2005. **Charles Woodland**

▲ Merchant Navy Pacific 35005 CANADIAN PACIFIC leaves Rothley for Loughborough on the Great Central Railway on 7 October 2006.   **Les Nixon**

▼ LMS Class 4F 0-6-0 No. 44422 with a Bury–Rawtenstall service at Irwell Vale on the East Lancashire Railway on 9 September 2006.   **George Allsop**

▶ Stanier Class 5MT 4-6-0 No. 45305 departs Quorn & Woodhouse on the Great Central Railway with the 09.15 Loughborough Central–Leicester North on 14 August 2004.
John Calton

▼ LMS Jubilee 4-6-0 No. 5690 LEANDER climbs towards Harpenden on 8 April 2006 with a Tyseley–St. Albans charter.
John Pink

54

LMS Princess Coronation Class 8P 4-6-2 No. 6233 DUCHESS OF SUTHERLAND passes Penmaenmawr with a charter train from Crewe to Holyhead (originally ex-Lincoln) on 16 September 2006.
Les Nixon

▲ LMS Princess Royal Pacific No. 6201 PRINCESS ELIZABETH passes Minshull Vernon with a Carlisle–Crewe charter on 29 August 2005. **Hugh Ballantyne**

▼ LMS Class 3F "Jinty" 0-6-0T No. 47279 approaches Oakworth on the Keighley & Worth Valley Railway with the 12.00 Keighley–Oxenhope on 1 May 2006. **Brian Dobbs**

▲ Stanier Class 8F 2-8-0 TCDD 45161 (ex-WD No. 522) on display at the TCDD open air Steam Loco Museum at Camlik, Turkey on 30 April 2007. **Peter Fox**

▼ Somerset & Dorset Joint Class 7F 2-8-0 53809 (double heading with West Country Pacific 34067) passes Blue Anchor Beach on the West Somerset Railway on 19 March 2006. **Charles Woodland**

▲ Caledonian Railway 0-4-4T No. 419 (BR 55189) passes Kinneil on the Bo'ness & Kinneil Railway with the 10.33 Bo'ness–Birkhill on 3 July 2004.
Ian Lothian

▼ LNER Class A4 Pacific No. 60007 SIR NIGEL GRESLEY is seen at Moorgates, south of Goathland on the North Yorkshire Moors Railway on 13 October 2006 with an afternoon Grosmont–Pickering service.
Paul Chancellor

▲ LNER Class V2 2-6-2 No. 4771 GREEN ARROW (BR 60800) at Colton with a King's Cross–Scarborough special on 12 August 2006.

▼ LNER Class K4 2-6-0 No. 61994 THE GREAT MARQUESS at Froghall on a visit to the Churnet Valley Railway on 14 January 2007.

Les Nixon (2)

▲ LNER Class B1 2-6-0 No. 61264 approaches Mauds Bridge on 15 April 2006 with the 07.54 King's Cross–Cleethorpes charter service. **Chris Booth**

▼ LNER design Class K1 2-6-0 No. 62005 at Green End on the North Yorkshire Moors Railway with the 16.50 Grosmont–Pickering on 22 March 2005. **Paul Chancellor**

▲ LNER Class D49 4-4-0 No. 246 MORAYSHIRE passes Kinneil on the Bo'ness & Kinneil Railway with the 12.20 Bo'ness–Birkhill on 15 April 2004. **Ian Lothian**

▼ LNER Class O4 2-8-0 No. 63601 (ex-GC) approaches Quorn on the Great Central Railway with a demonstration freight on 28 January 2007. **Les Nixon**

BR Standard Class 8P 4-6-2 No. 71000 DUKE OF GLOUCESTER departs from Carnforth after a servicing stop with a charter train from Carlisle to Preston via Settle on 3 April 2005.
Keith Satterly

▲ BR Standard Class 5MT 4-6-0 No. 73082 CAMELOT leaves Sheffield Park on the Bluebell Railway with a "Boat Train" charter on 3 November 2004. **Alan Barnes**

▼ BR Standard Class 4MT 4-6-0 No. 75029 runs round its train at Grosmont on the North Yorkshire Moors Railway on 20 September 2006. **John Calton**

▲ BR Standard Class 4MT 2-6-0 No. 76079 climbs past Moorgates on the North Yorkshire Moors Railway on 19 September 2006 with a lengthy train bound for Pickering. **George Allsop**

▼ BR Standard Class 2MT 2-6-0 No. 78019 at Woodthorpe on the Great Central Railway with the 14.15 Loughborough Central–Leicester on 6 March 2005. **John Pink**

▲ BR Standard Class 4MT 2-6-4T No. 80002 stands at Damens Loop with the 17.15 Keighley–Oxenhope on 1 May 2006. **Brian Dobbs**

▼ BR Standard Class 9F 2-10-0 No. 92214 passes Swanwick on the Midland Railway-Butterley's line with a Russ Hillier photo charter on 15 March 2003. **Alan Barnes**

CLASS Y1 4wT
Built: 1925–33. Sentinel geared loco. 24 built.
Boiler Pressure: 275 lb/sq. in. superheated.
Wheel Diameter: 2' 6".
Valve Gear: Rotary cam. Poppet valves.
Tractive Effort: 7260 lbf.
RA: 1.
Weight: 19.8 tons.
Cylinders: 6¾" x 9" (I).
BR Power Classification: 0F.

BR	LNER			
68153	59–8153		Middleton Railway	S 8837/1933

Also carried DEPARTMENTAL LOCOMOTIVE No. 54.

CLASS J69 0-6-0T
Built: 1890–1904. Holden GER class S56. 126 locomotives (including many rebuilt from J67).
Boiler Pressure: 180 lb/sq. in.
Wheel Diameter: 4' 0".
Valve Gear: Stephenson. Slide valves.
BR Power Classification: 2F.
Air brakes.
Weight: 42.45 tons.
Cylinders: 16½" x 22" (I).
Tractive Effort: 19 090 lbf.
RA: 3.

BR	LNER	GER		
68633	7087–8633	87	National Railway Museum	Stratford 1249/1904

CLASS J52 0-6-0ST
Built: 1897–1902. Ivatt GNR class J13. Many rebuilt from Stirling locomotives (built 1892–97).
Boiler Pressure: 170 lb/sq. in.
Wheel Diameter: 4' 8".
Valve Gear: Stephenson. Slide valves.
BR Power Classification: 3F.
Weight: 51.7 tons.
Cylinders: 18" x 26" (I).
Tractive Effort: 21 740 lbf.
RA: 5.

BR	LNER	GNR		
68846	4247–8846	1247	Locomotion: The NRM at Shildon	SS 4492/1899

CLASS J72 0-6-0T
Built: 1898–1925. Worsdell NER class E1. Further batch built 1949–51 by BR. 113 built.
Boiler Pressure: 140 lb/sq. in.
Wheel Diameter: 4' 1¼".
Valve Gear: Stephenson. Slide valves.
BR Power Classification: 2F.
Weight: 38.6 tons
Cylinders: 17" x 24" (I).
Tractive Effort: 16 760 lbf.
RA: 5.

69023–Departmental No. 59   NELPG, Hopetown, Darlington   Darlington 2151/1951

CLASS N2 0-6-2T
Built: 1920–29. Gresley GNR class N2. 107 built (69490–69596).
Boiler Pressure: 170 lb/sq. in. superheated.
Wheel Diameter: 5' 8", 3' 8".
Valve Gear: Stephenson. Piston valves.
BR Power Classification: 3MT.
Weight: 70.25 tons.
Cylinders: 19" x 26" (I).
Tractive Effort: 19 950 lbf.
RA: 6.

BR	LNER	GNR		
69523	4744–9523	1744	Great Central Railway	NBL 22600/1921

CLASS N7 0-6-2T
Built: 1915–28. Hill GER class L77. 134 built (69600–69733).
Boiler Pressure: 180 lb/sq. in. superheated.
Wheel Diameters: 4' 10", 3' 9".
Valve Gear: Walschaerts (inside). Piston valves.
RA: 5.
Dual brakes.
Weight: 61.8 tons.
Cylinders: 18" x 24" (I).
Tractive Effort: 20 510 lbf.
BR Power Classification: 3MT.

BR	LNER	GER			
69621	999E–7999–9621	999	"A.J. HILL"	North Norfolk Railway	Stratford 1924

CLASS X1   2-2-4T

Built: 1869 by NER as 2–2–2WT. Rebuilt 1892 to 4–2–2T and rebuilt as 2–cyl compound 2–2–4T and used for pulling inspections saloons. (NER Class 66).
Boiler Pressure: 175 lb/sq. in.   **Weight:** 44.95 tons.
Wheel Diameters: 3' 7", 5' 7¾", 3' 1¼".   **Cylinders:** 13" x 24" (hp) + 18½" x 20" (lp) (I).
Valve Gear: Stephenson. Slide valves.   **Tractive Effort:** 6390 lbf.

LNER	NER			
66	1478–66	AEROLITE	National Railway Museum	Gateshead 1869

NER 901 CLASS   2-4-0

Built: 1872–82. Fletcher design. 55 built.
Boiler Pressure: 160 lb/sq. in.   **Weight –Loco:** 39.7 tons.
Wheel Diameters: 4' 6", 7' 6".   **–Tender:** 29.9 tons.
Cylinders: 18" x 24" (I).   **Valve Gear:** Stephenson. Slide valves.
Tractive Effort: 12 590 lbf

LNER	NER		
910	910	Locomotion: The NRM at Shildon	Gateshead 1875

NER 1001 CLASS   0-6-0

Built: 1864–75. Bouch design for Stockton and Darlington Railway.
Boiler Pressure: 130 lb/sq. in.   **Weight –Loco:** 35 tons.
Wheel Diameter: 5' 0½".   **Cylinders:** 17" x 26" (I).
Valve Gear: Stephenson. Slide valves.   **Tractive Effort:** 13 720 lbf.

LNER	NER		
1275	1275	National Railway Museum	Darlington 708/1874

CLASS E5   2-4-0

Built: 1885. Tennant NER 1463 class. 20 built.
Boiler Pressure: 160 lb/sq. in.   **Weight –Loco:** 42.1 tons.
Wheel Diameters: 4' 6", 7' 0".   **–Tender:** 32.1 tons.
Cylinders: 18" x 24" (I).   **Valve Gear:** Stephenson. Slide valves.
Tractive Effort: 12 590 lbf.

LNER	NER		
1463	1463	Darlington North Road Museum (N)	Darlington 1885

CLASS D17   4-4-0

Built: 1893–7. Worsdell NER class M1 (later class M). 20 built.
Boiler Pressure: 160 lb/sq. in.   **Weight –Loco:** 52 tons.
Wheel Diameters: 3' 7¼", 7' 1¼".   **–Tender:** 41 tons.
Cylinders: 19" x 26" (I).   **Valve Gear:** Stephenson. Slide valves.
Tractive Effort: 14 970 lbf.   **RA:** 6.

LNER	NER		
1621	1621	National Railway Museum	Gateshead 1893

CLASS C2   "KLONDYKE"   4-4-2

Built: 1898–1903. H. A. Ivatt GNR class C1. 22 built.
Boiler Pressure: 170 lb/sq. in.   **Weight –Loco:** 62 tons.
Wheel Diameters: 3' 8", 6' 8", 3' 8".   **–Tender:** 42.1 tons.
Cylinders: 19" x 24" (O).   **Valve Gear:** Stephenson. Slide valves.
Tractive Effort: 15 650 lbf.   **RA:** 4.

LNER	GNR			
3990	990	HENRY OAKLEY	Bressingham Steam Museum (N)	Doncaster 769/1898

GNR 4-2-2
Built: 1870–95. Stirling design.
Boiler Pressure: 140 lb/sq. in.
Wheel Diameters: 3' 10", 8' 1", 4' 1".
Cylinders: 18" x 28" (O).
Tractive Effort: 11 130 lbf.
Weight –Loco: 38.5 tons.
    –Tender: 30 tons.
Valve Gear: Stephenson. Slide valves.

1   National Railway Museum   Doncaster 50/1870

STOCKTON & DARLINGTON RAILWAY 0-4-0
Built: 1825–6. George Stephenson design. 6 built.
Boiler Pressure: 50 lb/sq. in.
Wheel Diameter: 3' 11".
Cylinders: 9½" x 24" (O).
Weight –Loco: 6.5 tons.
    –Tender:
Tractive Effort: 2050 lbf.

1   LOCOMOTION   Darlington North Road Museum (N)   RS1/1825

STOCKTON & DARLINGTON RAILWAY 0-6-0
Built: 1845.
Boiler Pressure: 75 lb/sq. in.
Wheel Diameter: 4' 0".
Tractive Effort: 6700 lbf.
Weight –Loco: 6.5 tons.
Cylinders: 14½" x 24" (O).

25   DERWENT   Darlington North Road Museum (N)   Kitching 1845

1.4.1. NEW CONSTRUCTION
Three new construction projects are known about for ex-LNER and constituents' designs. The A1 TORNADO is the original new construction project and is nearing completion, whilst the two tank locos are at the early stages.

CLASS A1 4-6-2
Original Class Built: 1948–49. Peppercorn development of A2 with larger wheels. None of the class were preserved and this new locomotive is being constructed in Darlington.
Boiler Pressure: 250 lb/sq. in. superheated.
Wheel Diameters: 3' 2", 6' 8", 3' 8".
Cylinders: 19" x 26" (3).
Tractive Effort: 37 310 lbf.
RA: 9.
Weight –Loco: 105.2 tons.
    –Tender: 60.90 tons.
Valve Gear: Walschaerts. Piston valves.
BR Power Classification: 8P.
Air brakes.

60163 TORNADO   A1 Loco Trust, Hopetown, Darlington   Under construction

CLASS F5 2-4-2T
Original Class Built: 1903–09. Holden Great Eastern design for suburban use. The loco is eventually destined for the Epping–Ongar line. The original F5s finished at 67217, but two other locos, 67218/19 were reclassified to F5 from F6 in 1948.
Boiler Pressure: 180 lb/sq. in.
Wheel Diameters: 3' 9", 5' 4", 3' 9".
Valve Gear: Stephenson. Slide valves.
BR Power Classification: 1P.
Weight: 53.95 tons.
Cylinders: 17" x 24" (I).
Tractive Effort: 17 570 lbf.
RA: 3.

67218   Darlington   Under construction

CLASS G5 0-4-4T
Original Class Built: 1894–1901. Wilson Worsdell North Eastern Railway design for branch line and suburban use. 110 built (67240–67349). LNER 7306 was one of two members of the class withdrawn in 1950 and never received its allocated BR number 67306.
Boiler Pressure: 160 lb/sq. in.
Wheel Diameters: 3' 1¼", 5' 1¼".
Valve Gear: Stephenson. Slide valves.
BR Power Classification: 1P.
Weight: 54.20 tons.
Cylinders: 18" x 24" (I).
Tractive Effort: 17 200 lbf.
RA: 4.

67306   ?   Under construction

1.5. BRITISH RAILWAYS STANDARD STEAM LOCOMOTIVES

GENERAL
From 1951 onwards BR produced a series of standard steam locomotives under the jurisdiction of R.A. Riddles. Examples of most classes have been preserved, the exceptions being the Class 6MT "Clan" Pacifics, the Class 3MT 2-6-0s (77000 series), the Class 3MT 2-6-2Ts (82000 series) and the Class 2MT 2-6-2Ts (84000 series). Attempts are being made, however, to rectify two of the omissions.

Numbering System
Tender engines were numbered in the 70000 series and tank engines in the 80000 series, the exceptions being the class 9F 2-10-0s which were numbered in the 92000 series.

Classification System
British Railways standard steam locomotives were referred to by power classification like LMS locomotives. All locomotives were classed as "MT" denoting "mixed traffic" except for 71000 and the Class 9F 2-10-0s. The latter, although freight locos, were often used on passenger trains on summer Saturdays.

CLASS 7MT   BRITANNIA   4-6-2

Built: 1951–54. 55 built (70000–54).
Boiler Pressure: 250 lb/sq. in. (superheated).   **Weight –Loco:** 94 tons.
Wheel Diameters: 3′ 0″, 6′ 2″, 3′ 3½″.   **–Tender:** 49.15 tons.
Cylinders: 20″ x 28″ (O).   **Valve Gear:** Walschaerts. Piston valves.
Tractive Effort: 32 160 lbf.   **RA:** 7

x–Dual (air/vacuum) brakes.

70000 x	BRITANNIA	The Railway Age, Crewe	Crewe 1951
70013	OLIVER CROMWELL	Great Central Railway (N)	Crewe 1951

Note: 70013 will carry the number and name 70048 "THE TERRITORIAL ARMY 1908–2008" for a time from November 2007.

CLASS 8P   4-6-2

Built: 1953. 1 built.
Boiler Pressure: 250 lb/sq. in. (superheated).   **Weight –Loco:** 101.25 tons.
Wheel Diameters: 3′ 0″, 6′ 2″, 3′ 3½″.   **–Tender:** 53.7 tons.
Cylinders: 18″ x 28″ (3).   **Valve Gear:** British Caprotti (outside). Poppet valves.
Tractive Effort: 39 080 lbf.   **RA:** 8.
Dual (air/vacuum) brakes.

71000	DUKE OF GLOUCESTER	East Lancashire Railway	Crewe 1954

CLASS 5MT   4-6-0

Built: 1951–57. 172 built (73000–171).
Boiler Pressure: 225 lb/sq. in. (superheated).   **Weight –Loco:** 76 tons.
Wheel Diameters: 3′ 0″, 6′ 2″.   **–Tender:** 49.15 tons.
Cylinders: 19″ x 28″ (O).
Valve Gear: Walschaerts. Piston valves (*outside British Caprotti. Poppet valves).
Tractive Effort: 26 120 lbf.   **RA:** 5.

x–Dual (air/vacuum) brakes.

73050 x	"CITY OF PETERBOROUGH"	Nene Valley Railway	Derby 1954
73082	CAMELOT	Bluebell Railway	Derby 1955
73096		Mid Hants Railway	Derby 1955
73129 *		Midland Railway-Butterley	Derby 1956
73156		Great Central Railway	Doncaster 1956

CLASS 4MT   4-6-0

Built: 1951–57. 80 built (75000–79).
Boiler Pressure: 225 lb/sq. in. (superheated).   **Weight –Loco:** 67.9 tons.
Wheel Diameters: 3′ 0″, 5′ 8″.   **–Tender:** 42.15 tons.
Cylinders: 18″ x 28″ (O).   **Valve Gear:** Walschaerts. Piston valves.
Tractive Effort: 25 520 lbf.   **RA:** 4.

* Fitted with double chimney.

75014		Paignton & Dartmouth Railway	Swindon 1951
75027		Bluebell Railway	Swindon 1954
75029	"THE GREEN KNIGHT"	North Yorkshire Moors Railway	Swindon 1954
75069 *		Severn Valley Railway	Swindon 1955
75078 *		Keighley & Worth Valley Railway	Swindon 1956
75079 *		Mid Hants Railway	Swindon 1956

CLASS 4MT   2-6-0
Built: 1952–57. 115 built (76000–114).
Boiler Pressure: 225 lb/sq. in. (superheated).   **Weight –Loco:** 59.75 tons.
Wheel Diameters: 3' 0", 5' 3".   **–Tender:** 42.15 tons.
Cylinders: 17½" x 26" (O).   **Valve Gear:** Walschaerts. Piston valves.
Tractive Effort: 24 170 lbf.   **RA:** .

x–Dual (air/vacuum) brakes.

76017	Mid Hants Railway	Horwich 1953
76077	Gloucestershire-Warwickshire Railway	Horwich 1956
76079 x	East Lancashire Railway	Horwich 1957
76084	Ian Storey Engineering, Hepscott	Horwich 1957

CLASS 2MT   2-6-0
Built: 1952–56. 65 built (78000–64). These locomotives were almost identical to the Ivatt LMS Class 2MT 2-6-0s (46400–46527).
Boiler Pressure: 200 lb/sq. in. (superheated).   **Weight –Loco:** 49.25 tons.
Wheel Diameters: 3' 0", 5' 0".   **–Tender:** 36.85 tons.
Cylinders: 16½" x 24" (O).   **Valve Gear:** Walschaerts. Piston valves.
Tractive Effort: 18 510 lbf.   **RA:** .

78018	Darlington North Road Goods Shed	Darlington 1954
78019	Great Central Railway	Darlington 1954
78022	Keighley & Worth Valley Railway	Darlington 1954

CLASS 4MT   2-6-4T
Built: 1951–57. 155 built (80000–154).
Boiler Pressure: 225 lb/sq. in. (superheated).   **Weight:** 86.65 tons.
Wheel Diameters: 3' 0", 5' 8", 3' 0".   **Cylinders:** 18" x 28" (O).
Valve Gear: Walschaerts. Piston valves.   **Tractive Effort:** 25 520 lbf.
RA: 4.

80002	Keighley & Worth Valley Railway	Derby 1952
80064	Bluebell Railway	Brighton 1953
80072	Llangollen Railway	Brighton 1953
80078	Swanage Railway	Brighton 1954
80079	Severn Valley Railway	Brighton 1954
80080	Midland Railway-Butterley	Brighton 1954
80097	East Lancashire Railway	Brighton 1954
80098	Churnet Valley Railway	Brighton 1954
80100	Bluebell Railway	Brighton 1955
80104	Swanage Railway	Brighton 1955
80105	Bo'ness & Kinneil Railway	Brighton 1955
80135	North Yorkshire Moors Railway	Brighton 1956
80136	West Somerset Railway	Brighton 1956
80150	Barry Island Railway	Brighton 1956
80151	Bluebell Railway	Brighton 1957

Notes: 80098 is on loan from the Midland Railway-Butterley.
80136 is on loan from the Churnet Valley Railway.

CLASS 9F 2-10-0
Built: 1954–60. 251 built (92000–250). 92220 was the last steam locomotive to be built for British Railways.
Boiler Pressure: 250 lb/sq. in. (superheated). **Weight –Loco:** 86.7 tons.
Wheel Diameters: 3′ 0″, 5′ 0″.   **–Tender:** 52.5 tons.
Cylinders: 20″ x 28″ (O).   **Valve Gear:** Walschaerts. Piston valves.
Tractive Effort: 39 670 lbf.   **RA:** .

* Fitted with single chimney. All other surviving members of the class have double chimneys.

92134 *		The Railway Age, Crewe	Crewe 1957
92203	"BLACK PRINCE"	Gloucestershire-Warwickshire Railway	Swindon 1959
92207	"MORNING STAR"	North Dorset Railway, Shillingstone	Swindon 1959
92212		Mid Hants Railway	Swindon 1959
92214		Midland Railway-Butterley	Swindon 1959
92219		Midland Railway-Butterley	Swindon 1960
92220	EVENING STAR	National Railway Museum	Swindon 1960
92240		Bluebell Railway	Crewe 1958
92245		Barry Island Railway	Crewe 1958

1.5.1. NEW CONSTRUCTION

The Bluebell Railway are rebuilding a Class 2MT 2-6-0 for which they have no tender into a Class 2MT 2-6-2T. It will be numbered 84030, following on from the production series. The Severn Valley Railway are also building a Class 3MT 2-6-2T. Another group have started to build a "Clan" from scratch. Ten Clans were built, numbered 72000–09 and were all given names of Scottish Clans. Another 15 were ordered, but the order was cancelled and they were never built. The first five of these were to be for the Southern Region and were to be named HENGIST, HORSA, CANUTE, WILDFIRE and FIREBRAND respectively. The new locomotive will therefore be known as 72010 HENGIST.

CLASS 6MT   CLAN   4-6-2
New construction..
Built: 1951–52. 10 built (72000–09).
Boiler Pressure: 225 lb/sq. in. (superheated). **Weight –Loco:** 88.5 tons.
Wheel Diameters: 3′ 0″, 6′ 2″, 3′ 3½″.   **–Tender:** 49.15 tons.
Cylinders: 19½″ x 28″ (O).   **Valve Gear:** Walschaerts. Piston valves.
Tractive Effort: 27 520 lbf.   **RA:** 6

72010   HENGIST   North Dorset Railway, Shillingstone   Under construction

CLASS 3MT 2-6-2T
This class numbered 45 locos (82000–44) and were built 1952–55.
Boiler Pressure: 200 lb/sq. in. (superheated). **Weight:** 74.05 tons.
Wheel Diameters: 3′ 0″, 5′ 3″, 3′ 0″.   **Cylinders:** 17½″ x 26″ (O).
Valve Gear: Walschaerts. Piston valves.   **Tractive Effort:** 21 490 lbf.
RA: .

82045   Severn Valley Railway   Under construction

CLASS 2MT 2-6-2T
This locomotive was built in 1956 as Class 2MT 2-6-0 No. 78059 for which the Bluebell have no tender and is undergoing conversion.
Boiler Pressure: 200 lb/sq. in. (superheated). **Weight:** 66.25 tons.
Wheel Diameters: 3′ 0″, 5′ 0″, 3′ 0″.   **Cylinders:** 16½″ x 24″ (O).
Valve Gear: Walschaerts. Piston valves.   **Tractive Effort:** 18 510 lbf.
RA: .

84030   Bluebell Railway   Darlington 1956 reb. Bluebell

1.6. WAR DEPARTMENT LOCOMOTIVES

GENERAL

During the Second World War the War Department (WD) acquired and utilised a considerable number of steam locomotives. Upon the cessation of hostilities many of these locomotives were sold for further service both to industrial users and other railway administrations. The bulk of WD steam locomotives preserved date from this period. By 1952 many of the large wartime fleet of steam locomotives had been disposed of and those remaining were renumbered into a new series. From 1 April 1964 the WD became the Army Department of the Ministry of Defence with consequent renumbering taking place in 1968. The locomotives considered to be main line locomotives are included here. Also included in this section are the steam locomotives built to WD designs for industrial users.

CLASS WD   AUSTERITY   2-10-0

Built: 1943–45 by North British. 150 built. Many sold to overseas railways. 25 sold to British Railways in 1948 and numbered 90750–74.
Boiler Pressure: 225 lb/sq. in. superheated.   **Weight –Loco:** 78.3 tons.
Wheel Diameters: 3′ 2″, 4′ 8½″.   **–Tender:** 55.5 tons
Cylinders: 19″ x 28″ (O).   **Valve Gear:** Walschaerts. Piston valves.
Tractive Effort: 34 210 lbf.   **BR Power Classification:** 8F.

g Hellenic Railways (Greece) numbers.
n NS (Netherlands Railways) number.

WD	AD/Overseas			
3651–73651	600	GORDON	Severn Valley Railway	NBL 25437/1943
3652–73652	Lb951 g	"90775"	North Norfolk Railway	NBL 25438/1943
3656–73656	Lb955 g		Thessaloniki Depot, Greece (S)	NBL 25442/1943
3672–73672	Lb960 g	"DAME VERA LYNN"	North Yorkshire Moors Railway	NBL 25458/1944
3674–73674	Lb961 g		Achame Station, Athens, Greece	NBL 25460/1944
3677–73677	Lb962 g		Thessaloniki Depot, Greece	NBL 25463/1944
3682–73682	Lb964 g		Thessaloniki Depot, Greece	NBL 25468/1944
3684–73684	Lb966 g		Thessaloniki Depot, Greece (S)	NBL 25470/1944
3755–73755	5085 n	LONGMOOR	National Railway Museum, Utrecht	NBL 25541/1945

CLASS WD   AUSTERITY   2-8-0

Built: 1943–45 by North British & Vulcan Foundry. 935 built. Many sold to overseas railways. 200 were sold to LNER in 1946. These became LNER Nos. 3000–3199 and BR 90000–100, 90422–520. A further 533 were sold to British Railways in 1948 and numbered 90101–421, 90521–732.
Boiler Pressure: 225 lb/sq. in. superheated.   **Weight –Loco:** 70.25 tons.
Wheel Diameters: 3′ 2″, 4′ 8½″.   **–Tender:** 55.5 tons.
Cylinders: 19″ x 28″ (O).   **Valve Gear:** Walschaerts. Piston valves.
Tractive Effort: 34 210 lbf.   **BR Power Classification:** 8F.

This locomotive was purchased from the Swedish State Railways (SJ), it previously having seen service with Netherlands Railways (NS). It has been restored to BR condition.

NS	SJ	WD	Present		
4464	1931	79257	90733	Keighley & Worth Valley Railway	VF 5200/1945

CLASS 50550   0-6-0ST

Built: 1941–42. Eight locomotives were built to this design, all being intended for Stewarts and Lloyd Minerals. Only one was delivered with three being taken over by WD becoming 65–67. The other four went to other industrial users and two of these survive.
Boiler Pressure: 170 lb/sq. in.   **Weight:** 48.35 tons.
Wheel Diameter: 4′ 5″.   **Cylinders:** 18″ x 26″ (I).
Valve Gear: Stephenson. Slide valves.   **Tractive Effort:** 22 150 lbf.

WD	Present		
	Unnumbered	Rutland Railway Museum	HE 2411/1941
	GUNBY	Swindon & Cricklade Railway	HE 2413/1942
66–70066	S112 SPITFIRE	Embsay & Bolton Abbey Railway	HE 2414/1942

Note: 66 is receiving attention at Bryn Engineering, Pemberton, near Wigan.

CLASS WD AUSTERITY 0-6-0ST

Built: 1943–1953 for Ministry of Supply and War Department. 391 built. 75 bought by LNER and classified J94. Many others passed to industrial users. The design of this class was derived from the '50550' class of Hunslet locomotives (see above).
Boiler Pressure: 170 lb/sq. in.
Wheel Diameter: 4' 3".
Valve Gear: Stephenson. Slide valves.
BR Power Classification: 4F.
Weight: 48.25 tons.
Cylinders: 18" x 26" (I).
Tractive Effort: 23 870 lbf.
RA: 5.

§ Oil fired.
x Dual (air/vacuum) brakes.
† Rebuilt as 0-6-0 tender locomotive. Weights unknown.

BR	LNER	WD		
68078	8078	71463	Hope Farm, Sellindge	AB 2212/1946
68077	8077	71466	Spa Valley Railway	AB 2215/1946
WD/AD	Present			
71480	Unnumbered		Tyseley Locomotive Works, Birmingham	RSH 7289/1945
71499	Unnumbered		Bryn Engineering, Pemberton	HC 1776/1944
71505–118	BRUSSELS§		Keighley & Worth Valley Railway	HC 1782/1945
71515	68005		Embsay & Bolton Abbey Railway	RSH 7169/1944
71516	WELSH GUARDSMAN		Gwili Railway	RSH 7170/1944
71529–165	No. 15		Gloucestershire-Warwickshire Railway	AB 2183/1945
75006			Nene Valley Railway	HE 2855/1943
75008	SWIFTSURE		Strathspey Railway	HE 2857/1943
75015	48		Strathspey Railway	HE 2864/1943
75019–168	COAL PRODUCTS No.6		Rutland Railway Museum	HE 2868/1943
				rebuilt HE 3883/1962
75030	Unnumbered		Caledonian Railway	HE 2879/1943
75031–101	No. 17		Bo'ness & Kinneil Railway	HE 2880/1943
75041–107†	10 DOUGLAS		Mid Hants Railway	HE 2890/1943
				rebuilt HE 3882/1962
75050	No. 27 NORMAN		Hope Farm, Sellindge	RSH 7086/1943
75061	No. 9		Strathspey Railway	RSH 7097/1943
75062	49		Tanfield Railway	RSH 7098/1943
75080	NS 8811		Stoom Stichting Nederland, Rotterdam	HC 1737/1943
75091	ROBERT		The Railway Age, Crewe	HC 1752/1943
75105	WALKDEN		Ribble Steam Railway	HE 3155/1944
75113–132	SAPPER		South Devon Railway	HE 3163/1944
				rebuilt HE 3885/1964
75115	JULIA V		ZLSM, Simpelveld, Netherlands	HE 3165/1944
75118–134	WHELDALE		Embsay & Bolton Abbey Railway	HE 3168/1944
75130	No. 3180 ANTWERP		North Yorkshire Moors Railway	HE 3180/1944
75133–138	Unnumbered		Locomotion: The NRM at Shildon	HE 3183/1944
75141–139 x	68006		Peak Railway	HE 3192/1944
				rebuilt HE 3888/1964
75142–140	68012 BLACKIE		Lavender Line, Isfield	HE 3193/1944
				rebuilt HE 3887/1964
75158–144	THE DUKE 68012		Peak Railway	WB 2746/1944
75161	Unnumbered		Caledonian Railway	WB 2879/1944
75170	Unnumbered		Cefn Coed Colliery Museum	WB 2758/1946
75171–147	Unnumbered		Caledonian Railway	WB 2759/1944
75178	Unnumbered		Bodmin Steam Railway	WB 2766/1945
75186–150	ROYAL ENGINEER		Peak Railway	RSH 7136/1944
				rebuilt HE 3892/1969
75189–152	No. 8		Pontypool & Blaenavon Railway	RSH 7139/1944
				rebuilt HE 3880/1962
75254–175	No. 7		Bo'ness & Kinneil Railway	WB 2777/1945
75256	No. 20 TANFIELD		Tanfield Railway	WB 2779/1945
75282–181	Unnumbered		Gwili Railway	VF 5272/1945
				rebuilt HE 3879/1961
75300	3.55		Tunisville, Tunisia	VF 5290/1945
75319	72		Colne Valley Railway	VF 5309/1945
190–90			Colne Valley Railway	HE 3790/1952

191–91	No. 23 HOLMAN F. STEPHENS	Kent & East Sussex Railway	HE 3791/1952
192–92	WAGGONER	Isle of Wight Steam Railway	HE 3792/1953
193–93	Unnumbered	Ribble Steam Railway	HE 3793/1953
194–94	CUMBRIA	Lakeside & Haverthwaite Railway	HE 3794/1953
196	68011 ERROL LONSDALE	South Devon Railway	HE 3796/1953
197	No. 25 NORTHIAM	Kent & East Sussex Railway	HE 3797/1953
198–98	ROYAL ENGINEER	Isle of Wight Steam Railway	HE 3798/1953
200	No. 24 ROLVENDEN	Kent & East Sussex Railway	HE 3800/1953

Notes:

75115 carries worksplate 3164/1944 and 196 carries worksplate 3799/1953.
75129 is currently receiving attention at Rye Farm, Wishaw.

In addition to the 391 locomotives built for the Ministry of Supply and the War Department, a further 93 were built for industrial users, those that survive being:

No. 60	Strathspey Railway	HE 3686/1949
WHISTON	Foxfield Railway	HE 3694/1950
IR	Ribble Steam Railway	HE 3696/1950
11 REPULSE	Lakeside & Haverthwaite Railway	HE 3698/1950
NORMA	Oswestry Cycle & Railway Museum	HE 3770/1952
8 SIR ROBERT PEEL	Embsay & Bolton Abbey Railway	HE 3776/1952
"68030"	Churnet Valley Railway	HE 3777/1952
THOMAS 1	Mid Hants Railway	HE 3781/1952
No. 69	Embsay & Bolton Abbey Railway	HE 3785/1953
MONCKTON No.1	Embsay & Bolton Abbey Railway	HE 3788/1953
WILBERT	Dean Forest Railway	HE 3806/1953
3809	Weardale Railway	HE 3809/1954
GLENDOWER	South Devon Railway	HE 3810/1954
68019	Bo'ness & Kinneil Railway	HE 3818/1954
WARRIOR	Dean Forest Railway	HE 3823/1954
LNER 8009	Battlefield Railway	HE 3825/1954
Unnumbered	Swansea Vale Railway	HE 3829/1955
No. 5	Bo'ness & Kinneil Railway	HE 3837/1955
WIMBLEBURY	Foxfield Railway	HE 3839/1956
PAMELA	Barry Island Railway	HE 3840/1956
Unnumbered	Appleby-Frodingham RPS, Scunthorpe	HE 3846/1956
JUNO	Buckinghamshire Railway Centre	HE 3850/1958
CADLEY HILL No. 1	Snibston Discovery Park	HE 3851/1962
28	Rutland Railway Museum	HE 3889/1964
66	Buckinghamshire Railway Centre	HE 3890/1964

SHROPSHIRE & MONTGOMERY RLY   0-4-2WT

Built: 1893 by Alfred Dodman & Company for William Birkitt. Sold to Shropshire & Montgomery Railway in 1911 when it was rebuilt from 2-2-2WT.
Boiler Pressure: ? lbf/sq. in.   **Weight:** 5.5 tons.
Wheel Diameters: 2' 3", 2'3".   **Cylinders:** 4" x 9" (I).

Note: This locomotive together with three others (LNWR coal engines) became BR (WR) stock in 1950 when this line was nationalised. The locomotives were then withdrawn.

1	GAZELLE	Kent & East Sussex Railway	Dodman 1893

1.7. UNITED STATES ARMY TRANSPORTATION CORPS STEAM LOCOMOTIVES

CLASS S160   2-8-0

Built: 1942–45 by various American builders. It is estimated that 2120 were built. Many saw use on Britain's railways during the Second World War. Post war many were sold to overseas railway administrations. Only those currently in Great Britain are shown here.
Boiler Pressure: 225 lb/sq. in. superheated.   **Weight –Loco:** 73 tons.
Wheel Diameters: 3' 2", 4' 9".   **–Tender:** 52.1 tons.
Cylinders: 19" x 26" (O).   **Tractive Effort:** 31490 lbf.
Valve Gear: Walschaerts. Piston valves.

USATC Overseas Railway

1631	MAV (Hungarian Railways) 411.388	Nottingham Heritage Centre	AL 70284/1942
2253	PKP (Polish Railways) Ty203-288	North Yorkshire Moors Railway	BLW 69496/1943
2364	MAV (Hungarian Railways) 411.337	East Lancashire Railway	BLW 69621/1943
3278	FS (Italian State Railways) 736-073	Mid Hants Railway	AL 71533/1944
5197	Chinese State Railways SD6.463	Churnet Valley Railway	Lima 8856/1945
5820	PKP (Polish Railways) Ty203-474	Keighley & Worth Valley Railway	Lima 8758/1945
6046	MAV (Hungarian Railways) 411.144	Churnet Valley Railway	BLW 72080/1945

Note: 3278 passed to Hellenic Railways (Greece) and numbered 575. Carries name "FRANKLYN D. ROOSEVELT".

See also SR section for USA 0-6-0Ts.

1.8. REPLICA STEAM LOCOMOTIVES
GENERAL

Details are included of locomotives which would have been included in the foregoing had the locomotive they replicate survived. Locomotives are listed under the heading of the Railway/Manufacturer they replicate. Only locomotives which work, previously worked or can be made to work are included.

STOCKTON & DARLINGTON RAILWAY   0-4-0
Built: 1975.
Boiler Pressure: 50 lb/sq. in.   **Weight –Loco:** 6.5 tons.
Wheel Diameter: 3' 11"   **–Tender:**
Cylinders: 9½" x 24" (O).   **Tractive Effort:** 1960 lbf.

LOCOMOTION   North of England Open Air Museum   Loco. Ent.No. 1/1975

LIVERPOOL & MANCHESTER RAILWAY   0-2-2
Built: 1934/79.
Boiler Pressure: 50 lb/sq. in.   **Weight –Loco:** 4.5 tons.
Wheel Diameters: 4' 8½", 2'6".   **–Tender:** 5.2 tons
Cylinders: 8" x 17" (O).   **Tractive Effort:** 820 lbf.

ROCKET   National Railway Museum   RS 4089/1934
ROCKET   National Railway Museum   Loco. Ent.No. 2/1979

LIVERPOOL & MANCHESTER RAILWAY   0-2-2
Built: 1979.
Boiler Pressure: 50 lb/sq. in.   **Weight –Loco:** 4.5 tons.
Wheel Diameter: 4' 6"   **–Tender:** 5.2 tons.
Cylinders: 7" x 18" (O).   **Tractive Effort:** 690 lbf.

SANS PAREIL   Locomotion: The NRM at Shildon   Shildon/1980

LIVERPOOL & MANCHESTER RAILWAY   2-2-0
Built: 1992.
Boiler Pressure: **Weight –Loco:**
Wheel Diameters: **–Tender:**
Cylinders: **Tractive Effort:**

PLANET   Greater Manchester Museum of Science & Industry   Manch 1992

BRAITHWAITE & ERICCSON & COMPANY   0-4-0
Built: 1929 (using parts from 1829 original)/1980.
Boiler Pressure: **Weight –Loco:**
Wheel Diameter: **–Tender:**
Cylinders: **Tractive Effort:**

NOVELTY   Greater Manchester Museum of Science & Industry (N)   Science Museum 1929
NOVELTY   Swedish Railway Museum, Gayle, Sweden   Loco. Ent. No. 3 1980

GREAT WESTERN RAILWAY   2-2-2
Built: 1925 (using parts from 1837 original).   **Gauge:** 7' 0¼".
Boiler Pressure: 90 lb/sq. in.   **Weight –Loco:** 23.35 tons.
Wheel Diameters: 4' 0", 7' 0",4' 0".   **–Tender:** 6.5 tons.
Cylinders: 16" x 16" (I)   **Tractive Effort:** 3730 lbf.

NORTH STAR   Steam – Museum of the Great Western Railway (N)   Swindon 1925

GREAT WESTERN RAILWAY 2-2-2
Built:
Boiler Pressure: 50 lb/sq. in.
Wheel Diameters: 4' 0", 7' 0", 4' 0".
Cylinders: 15" x 18" (I).
Gauge: 7' 0¼".
Weight –Loco: 23.35 tons
   –Tender: 6.5 tons
Tractive Effort: 2050 lbf.

FIRE FLY   Didcot Railway Centre

GREAT WESTERN RAILWAY 4-2-2
Built: 1985.
Boiler Pressure:
Wheel Diameters: 4' 6", 8' 0", 4' 6".
Cylinders: 18" x 24" (I).
Gauge: 7' 0¼".
Weight –Loco: 35.5 tons
   –Tender:
Tractive Effort: 7920 lbf.

IRON DUKE   National Railway Museum   Resco 1985

1.9. OVERSEAS STEAM LOCOMOTIVES PRESERVED IN GREAT BRITAIN

Note: All locomotives are standard gauge unless shown otherwise. Only locomotives of minimum 750 mm gauge are listed. They must either be locomotives of foreign main line railways, or they must be preserved at a site which was once part of the British national network. Dimensions are mainly in metric units in this section.

Full details were not to hand for all locomotives in this section at the time of going to press.

1.9.1. ANTIGUA

UNCLASSIFIED   0-6-2T

Built: 1927 for Antigua Sugar Factory Railway.   **Gauge:** 760 mm (2'6").
Boiler Pressure: **Weight:**
Wheel Diameter: **Cylinders:**
Valve Gear: **Tractive Effort:**

Ant	Present		
?	JOAN "No. 12"	Welshpool & Llanfair Railway	KS 4404/1927

1.9.2. AUSTRIA

CLASS 699   0-8-0T

Built: 1944. German Wehrmacht design. Purchased by Salzkammergut Lokalbahn and later worked on Steiermärkische Landesbahnen and Zillertalbahn.
Gauge: 760 mm (2'6").
Boiler Pressure: 12 bar.   **Weight:** 25 tonnes.
Wheel Diameter: 630 mm   **Cylinders:** 330 x 310 mm (O)
Valve Gear: Walschaerts   **Tractive Effort:**

SKLB StLB/ZB	Present		
19 699.01	"SIR DREFALDWYN No. 10"	Welshpool & Llanfair Railway	FB 2855/1944

1.9.3. CHINA

CLASS KF   4-8-4

Built: 1935 for Chinese National Railways.
Boiler Pressure: **Weight –Loco:**
Wheel Diameters: **–Tender:**
Cylinders: (O).   **Valve Gear:** Walschaerts. Piston valves.
Tractive Effort:
607   National Railway Museum   VF 4674/1935

1.9.4. DENMARK

CLASS WS   0-4-0T

Built: 1895 .
Boiler Pressure: **Weight:**
Wheel Diameter: **Cylinders:**
Valve Gear: **Tractive Effort:**
DSB
385   Middleton Railway   Hartmann 2110/1895

CLASS F 0-6-0T
Built: 1944.
Boiler Pressure:
Wheel Diameter:
Valve Gear: Allan.
Weight:
Cylinders:
Tractive Effort:
DSB
656 Nene Valley Railway Frichs 360/1949

CLASS E 4-6-2
Built: 1950.
Boiler Pressure:
Wheel Diameters:
Valve Gear:
Tractive Effort:
Weight:
Cylinders:.

DSB
996 Railworld, Peterborough Frichs 415/1950

1.9.5. FINLAND
Gauge: 1524 mm (5'0").

CLASS Vr1 0-6-0T
Built: 1913–27. 43 built.
Boiler Pressure: 12 kp/cm^2
Wheel Diameter: 1270 mm
Valve Gear:
Weight: 44.8 tonnes.
Cylinders: 430 x 550 mm.
Tractive Effort: 6250 kp
VR *Present*
792 792 "HEN" Enfield Timber Company TK 373/1927
794 Epping & Ongar Railway TK 350/1925
799 Hope Farm, Sellindge TK 355/1925

CLASS Hr1 4-6-2
Built: 1957–57. 42 built.
Boiler Pressure: 12 kp/cm^2.
Wheel Diameters: 860, 1900 mm
Cylinders: 590 x 650 mm
Tractive Effort: 11610 kp
Weight –Loco: 55.2 tonnes.
–Tender: 33.3 tonnes.
Valve Gear:
VR
1008 Epping & Ongar Railway LO 157/1948
1016 Long & Somerville, Southbury, London TK 946/1948

CLASS Tr1 2-8-2
Built: 1940–57. 67 built.
Boiler Pressure: 15 kp/cm^2·
Wheel Diameters: 960, 1600, 1120 mm
Cylinders: 610 x 700 mm.
Tractive Effort: 15900 kp.
Weight –Loco: 95 tonnes.
–Tender: 62 tonnes.
Valve Gear:
VR
1060 Epping & Ongar Railway LO 172/1954
1077 Steam Traction, Acton, Suffolk Jung 11787/1953

CLASS Tk3 2-8-0
Built: 1927–53. (161 built).
Boiler Pressure: 14 kp/cm^2.
Wheel Diameters: 960, 1600 mm
Cylinders: 460 x 630 mm
Tractive Effort: 9550 kp.
Weight –Loco: 51.8 tonnes.
–Tender: 26.4 tonnes.
Valve Gear:

VR			
1103		Spirit of the West American Theme Park	LO 141/1945
1134		Epping & Ongar Railway	TK 531/1946
1144		Steam Traction, Acton, Suffolk	TK 571/1948
1151		Epping & Ongar Railway	TK 397/1949
1157		HLPG, Binbrook Trading Estate	Frichs 403/1949

1.9.6. FRANCE

CLASS 230.D   4-6-0

Built: 1911. Du Bosquet–De Glehn 4-cylinder compound.
Boiler Pressure: **Weight –Loco:**
Wheel Diameters: **–Tender:**
Cylinders: **Valve Gear:**
Tractive Effort:

Nord	SNCF		
628	230.D.116	Nene Valley Railway	Hen 10745/1911

1.9.7. GERMANY

CLASS 64   2-6-2T

Built: 1928.
Boiler Pressure: 14 bar superheated.   **Weight:** 71 tons.
Wheel Diameters: 850, 1500, 850 mm.   **Cylinders:** (O) 500 x 660 mm.
Valve Gear:
Tractive Effort: .   **Max. Speed:** 90 km/h.

DB	DB (later)		
64.305	064 305-6	Nene Valley Railway	Krupp 1308/1934

1.9.8. NORWAY

CLASS 21   2-6-0

Built:
Boiler Pressure: xxx lb/sq. in.   **Weight –Loco:** tons.
Wheel Diameters: **–Tender:** tons.
Cylinders: **Valve Gear:**
Tractive Effort:

NSB	Present		
376		Kent & East Sussex Railway	NOHAB 1163/1919
377	377 "KING HAAKON 7"	Bressingham Steam Museum	NOHAB 1164/1919

CLASS 63a   2-6-0

Built: 1942–45. Standard Deutsche Reichsbahn Kriegslok (war loco).
Boiler Pressure: 16 bar superheated.   **Weight:** 89 tons.
Wheel Diameters: 850, 1400 mm.   **Cylinders:** 600 x 660 (O).
Valve Gear: **Tractive Effort:** lbf.

DR	NSB	Present		
52.5865	5865	5865 "PEER GYNT"	Bressingham Steam Museum	Schichau Unknown/1944

1.9.9. PAKISTAN

CLASS   4-4-0

Built: 1911 for PWR.
Boiler Pressure: **Gauge:** 1676 mm (5'6").
Wheel Diameter: **Weight:**
Valve Gear: Stephenson. Slide valves.   **Cylinders:**
                                                Tractive Effort:

| 3157 | Greater Manchester Museum of Science & Industry | VF 3064/1911 |

1.9.10. POLAND
CLASS Tkh "FERRUM 47" 0-6-0T
Built: 1947–60 for Polish industrial lines.
Boiler Pressure: 16 bar. **Weight**: 66.0 tonnes.
Wheel Diameter: **Cylinders**: 550 x 550 (O).
Valve Gear: **Tractive Effort**:

TKh.2871		Spa Valley Railway	Chr 2871/1951
TKh.2944	"HOTSPUR"	Spa Valley Railway	Chr 2944/1952
TKh.3135	"SPARTAN"	Spa Valley Railway	Chr 3135/1953
TKh.3138	"HUTNIK"	Appleby-Frodingham RPS	Chr 3138/1954
TKh.4015	"KAROL"	Avon Valley Railway	Chr 4015/1954
TKh.5374	"VANGUARD"	Northampton & Lamport Railway	Chr 5374/1959
TKh.7646	"Northamptonshire"	Northampton & Lamport Railway	Chr 3112/1952

CLASS Tkh "SLASK" 0-8-0T
Built: 1942–63 for Polish industrial lines. Henschel design.
Boiler Pressure: **Weight**:
Wheel Diameter: **Cylinders**:
Valve Gear: Walschaerts. **Tractive Effort**:

TKp.5485 Nene Valley Railway Chr 5485/1961

CLASS Ty2 KRIEGSLOK 2-10-0
Built: 1942–45. Standard Deutsche Reichsbahn Kriegslok (war loco).
Boiler Pressure: 16 bar superheated. **Weight**: 89 tons.
Wheel Diameters: 850, 1400 mm. **Cylinders**: 600 x 660 (O).
Valve Gear: Walschaerts.
Tractive Effort: lbf.

DR	Ind		
52.7173	Ty2.7173	Nene Valley Railway	Florids 16626/1943

1.9.11. ROMANIA
CLASS 764 0-8-0T
Built: 1954. **Gauge**: 760 mm (2'6").
Boiler Pressure: **Weight**: 24.8 tonnes.
Wheel Diameter: **Cylinders**:
Valve Gear: **Tractive Effort**:

CFI	Present		
764.423	INV 60006	Welshpool & Llanfair Railway	Resita 1679/1954
764.425	No. 19	Welshpool & Llanfair Railway	Resita ?/1954

1.9.12. SIERRA LEONE
Built: 1944. **Gauge**: 760 mm (2'6").
Boiler Pressure: **Weight**:
Wheel Diameter: **Cylinders**:
Valve Gear: **Tractive Effort**:

SLR	Present		
85	"No. 14"	Welshpool & Llanfair Railway	HE 3815/1954

1.9.13. SOUTH AFRICA

CLASS 7   4-8-0
Built: 1896.   **Gauge:** 1067 mm (3'6").
Boiler Pressure: **Weight:**
Wheel Diameter: **Cylinders:**
Valve Gear: **Tractive Effort:**
SAR
390   Rye Farm, Wishaw   SS 4150/1896

Note: SAR 390 was latterly used by the Zambesi Sawmills Railway.

CLASS GL   BEYER-GARRATT   4-8-2+2-8-4
Built: .   **Gauge:** 1067 mm (3'6").
Boiler Pressure: **Weight:**
Wheel Diameter: **Cylinders:**
Valve Gear: **Tractive Effort:**
SAR
2352   Greater Manchester Museum of Science & Industry   BP 6639/1930

CLASS 15F   4-8-2
Built: .   **Gauge:** 1067 mm (3'6").
Boiler Pressure: **Weight:** 179 tonne.
Wheel Diameter: **Cylinders:**
Valve Gear: **Tractive Effort:**
SAR
3007   Glasgow Museums Resource Centre, South Nitshill   NBL 25546/1944

CLASS 25NC   4-8-4
Built: .1953–54.   **Gauge:** 1067 mm (3'6").
Boiler Pressure: **Weight:**
Wheel Diameter: **Cylinders:**
Valve Gear: **Tractive Effort:**
SAR
3405   CITY OF BLOEMFONTEIN   Buckinghamshire Railway Centre   NBL 27291/1953

CLASS GMAM   BEYER-GARRATT   4-8-2+2-8-4
Built: .1953–58.   **Gauge:** 1067 mm (3'6").
Boiler Pressure: **Weight:**
Wheel Diameter: **Cylinders:**
Valve Gear: **Tractive Effort:**
SAR
4112   SPRINGBOK   Summerlee Heritage Park, Coatbridge   NBL 27770/1957

1.9.14. SWEDEN

CLASS S   2-6-2T
Built: .
Boiler Pressure: **Weight:**
Wheel Diameter: **Cylinders:**
Valve Gear: **Tractive Effort:**
SJ
1178   Nene Valley Railway   Motala 516/1904

CLASS Si 2-6-4T
Built:
Boiler Pressure: **Weight:**
Wheel Diameter: **Cylinders:**
Valve Gear: **Tractive Effort:**
SJ
1928 "ALLEN GLADDEN" Spa Valley Railway NOHAB 2229/1953

CLASS B 4-6-0
Built: 1954.
Boiler Pressure: **Weight:**
Wheel Diameter: **Cylinders:**
Valve Gear: **Tractive Effort:**
SJ
1313 Stephenson Railway Museum Motala 586/1917
1697 "SWJB 101" Nene Valley Railway NOHAB 2082/1944

1.9.15. TASMANIA

CLASS ?? 4-8-0
Built: 1951.
Boiler Pressure: **Gauge:** 1067 mm (3'6").
Wheel Diameter: **Weight:**
Valve Gear: **Cylinders:**
Tractive Effort:
TGR
M2 Tanfield Railway RSHN 7430/1951

1.9.16. TRINIDAD

CLASS S 2-6-2T
Built: .
Boiler Pressure: **Weight:**
Wheel Diameter: **Cylinders:**
Valve Gear: **Tractive Effort:**
USM
– PICTON Middleton Railway HE 1450/1927

1.9.17. YUGOSLAVIA

CLASS 62 0-6-0T
Built: This class is a copy of the USATC 0-6-0T (see section 1.2).
Boiler Pressure: **Weight:**
Wheel Diameter: **Cylinders:**
Valve Gear: **Tractive Effort:**
Ind Present
62-521 Mid Hants Railway Djora 1954/0-6-0T
62-669 "30075" East Somerset Railway Brod 669/1960

RAILWAY ABBREVIATIONS

ANT	Antigua Sugar Factory Railway	SJ	Statens Järnväger
CNR	Canadian National Railway	SKLB	Salzkammergut Lokalbahn
DB	Deutsche Bundesbahn	SLR	Sierra Leone Railway
DSB	Danske Statsbaner	SNCF	Société Nationale des Chemins de Fer Francais
Ind	Industrial Railway	StLB	Steirmarkische Landesbahnen
LAS	? (Poland)	TGR	Tasmanian Government Railways
NSB	Norges Statsbaner	USM	Usine Sainte Madeleine Sugar Mill Railway
SAR	South African Railways	VR	Valtion Rautatiet

2. DIESEL LOCOMOTIVES
GENERAL

It was not until the mid-1950s that diesel locomotives appeared in great numbers. However, during the 1930s, particularly on the LMS with diesel shunting locomotives, several small building programmes were authorised. A few locomotives survive from this period along with several others built post war. During the Second World War a large proportion of LMS locomotives were transferred to the war department whom also authorised the construction of further orders. Many of these were shipped across the English Channel and were subsequently lost in action. It is, however, possible some still remain undiscovered by the authors. Notification of these would be gratefully appreciated. Those known to survive are listed in Section 2.1.

From the mid-1950s the re-equipment of the railway network began in earnest and vast numbers of new, mainly diesel locomotives were constructed by British Railways and private contractors. Those from this period now considered to be preserved can be found in section 2.3. A few examples from this period are still in service with main line and industrial users, details of these can be found in other Platform 5 books.

NOTES
Wheel Arrangement

The Whyte notation is used for diesel shunters with coupled driving wheels (see steam section). For other shunting and main-line diesel and electric locomotives, the system whereby the number of driven axles on a bogie or frame is denoted by a letter (A = 1, B = 2, C = 3 etc) and the number of undriven axles is denoted by a number is used. The letter "o" after a letter indicates that each axle is individually powered and a + sign indicates that the bogies are intercoupled.

Dimensions

SI units are generally used. Imperial units are sometimes given in parentheses.

Tractive Effort

Continuous and maximum tractive efforts are generally quoted for vehicles with electric transmission.

Brakes

Locomotives are assumed to have train vacuum brakes unless otherwise stated.

Numbering Systems

Prior to nationalisation each railway company allocated locomotive numbers in accordance with its own policy. However, after nationalisation in 1948 a common system was devised and internal combustion locomotives were allocated five figure numbers in the series 10000–19999.

Diesel locomotives built prior to nationalisation or to pre-nationalisation designs are arranged generally in order of the 1948 numbers with those withdrawn before 1948 listed at the beginning of each section. In 1957 a new numbering scheme was introduced for locomotives built to British Railways specifications and details of this are given in the introduction to the British Railways section.

2.1. LONDON MIDLAND & SCOTTISH RAILWAY

DIESEL MECHANICAL 0-4-0
Built: 1934 by English Electric at Preston Works for Drewry Car Company.
Engine: Allan 8RS18 of 119 kW (160 h.p.) at 1200 r.p.m. (Now fitted with a Gardner 6L3 of 114 kW (153 h.p.).
Transmission: Mechanical. Wilson four-speed gearbox driving a rear jackshaft.
Maximum Tractive Effort: 50 kN (15300 lbf).   **Weight:** 25.8 tonnes.
Maximum Speed: 12 m.p.h.   **Wheel Diameter:** 914 mm.
No train brakes.

LMS	WD	AD			
7050	224–70224–846	240	"RORKE'S DRIFT" National Railway Museum	DC 2047/EE/DK 847 1934

DIESEL MECHANICAL 0-6-0
Built: 1932 by Hunslet Engine Company (taken into stock 1933).
Engine: MAN 112 kW (150 h.p.) at 900 r.p.m. (Now fitted with a Maclaren/Ricardo 98 kW engine).
Transmission: Mechanical. Hunslet constant mesh four-speed gearbox.
Maximum Tractive Effort: **Weight:** 21.7 tonnes.
Maximum Speed: 30 m.p.h.   **Wheel Diameter:** 914 mm.
Built without train brakes, but vacuum brakes now fitted.

7401–7051   "JOHN ALCOCK"   Middleton Railway   HE 1697/1933

DIESEL ELECTRIC 0-6-0
Built: 1935 by English Electric. 11 built, 3 taken over by BR as 12000–2 (LMS 7074/6/9). Scrapped 1956–62. Others sold to WD for use in France.
Engine: English Electric 6K of 261 kW (350 h.p.) at 675 r.p.m.
Transmission: Electric. Two axle-hung traction motors with a single reduction drive.
Maximum Tractive Effort: 147 kN (33000 lbf).   **Weight:** 52 tonnes.
Maximum Speed: 30 m.p.h.   **Wheel Diameter:** 1232 mm.
No train brakes.

LMS	WD		
7069	18	Gloucestershire-Warwickshire Railway	EE/HL 3841 1935

DIESEL ELECTRIC 0-6-0
Built: 1939–42 at Derby. 40 built, 30 taken over by BR as 12003–32 (LMS 7080–99, 7110–9). Scrapped 1964–7. Others sold to WD for use in Italy and Egypt.
Engine: English Electric 6KT of 261 kW (350 h.p.) at 680 r.p.m.
Transmission: Electric. One traction motor with jackshaft drive.
Maximum Tractive Effort: 147 kN (33000 lbf).   **Weight:** 56 tonnes.
Maximum Speed: 20 m.p.h.   **Wheel Diameter:** 1295 mm.
Built without train brakes, but air brakes now fitted.

LMS	WD	FS	Present		
7103	52–70052	700.001	Unnumbered	Piedmont Rail Museum, Torino, Italy	Derby 1941
7106	55–70055	700.003	Unnumbered	LFI, Arezzostia Pescaiola, Italy	Derby 1941

BR CLASS 11   DIESEL ELECTRIC   0-6-0

Built: 1945–53. LMS design. 120 built. The first order for 20 was for the WD, 14 being delivered as 260–8/70269–73, the balance passing to the LMS as 7120–7125 (BR 12033–38). 260–8 were renumbered 70260–8 and 70260–9 were sold to the NS (Netherlands Railways) in 1946.
Engine: English Electric 6KT of 261 kW (350 h.p.) at 680 r.p.m.
Transmission: Electric. Two EE 506 axle hung traction motors.
Maximum Tractive Effort: 156 kN (35000 lbf).
Continuous Tractive Effort: 49.4 kN (11100 lbf) at 8.8 m.p.h.
Wheel Diameter: 1372 mm.   **Weight:** 56 tonnes.
Power at Rail: 183 kW (245 h.p.).   **Maximum Speed:** 20 m.p.h.

No train brakes except 70272, 12099 and 12131 which have since been fitted with vacuum brakes and 12083 which has since been fitted with air brakes.

BR	WD	AD	Present		
	70269		508§	Netherlands National Railway Museum, Blerik Yard Storage Shed	Derby 1944
	70272–878	601	7120	Lakeside & Haverthwaite Railway	Derby 1944
12049				Mid Hants Railway	Derby 1948
12052			MP228	Caledonian Railway	Derby 1949
12061			4	Peak Railway	Derby 1949
12077				Midland Railway-Butterley	Derby 1950
12083				Battlefield Railway	Derby 1950
12093			MP229	Caledonian Railway	Derby 1951
12099				Severn Valley Railway	Derby 1952
12131				North Norfolk Railway	Darlington 1952

§ NS number.

2.2. SOUTHERN RAILWAY

BR CLASS 12   DIESEL ELECTRIC   0-6-0

Built: 1949–52. Based on pre-war LMS design. 26 built.
Engine: English Electric 6KT of 350 h.p. at 680 r.p.m.
Transmission: Electric. Two EE 506A axle-hung traction motors.
Maximum Tractive Effort: 107 kN (24000 lbf).
Continuous Tractive Effort: 36 kN (8000 lbf) at 10.2 m.p.h.
Wheel Diameter: 1370 mm.   **Weight:** 45 tonnes.
Power at Rail: 163 kW (218 h.p.).   **Maximum Speed:** 27 m.p.h.
No train brakes.

15224	Spa Valley Railway	Ashford 1949

2.3. BRITISH RAILWAYS

Numbering System

In 1957 British Railways introduced a new numbering system which applied to all diesel locomotives except those built to pre-nationalisation designs. Each locomotive was allocated a number of up to four digits prefixed with a "D". Diesel electric shunters already built numbered in the 13xxx series had the "1" replaced by a "D". Diesel mechanical shunters already built numbered in the 11xxx series were allocated numbers in the D2xxx series.

When all steam locomotives had been withdrawn, the prefix letter was officially eliminated from the number of diesel locomotives, although it continued to be carried on many of them. For this reason, no attempt is made to distinguish between those locomotives which did or did not have the "D" prefix removed. Similarly, in preservation, no distinction is made between locomotives which do or do not carry a "D" prefix at present. In 1968 British Railways introduced a numerical classification code for diesel & electric locomotives.

With the introduction of modern communications each locomotive was allocated a two digit class number followed by a three digit serial number. These started to be applied in 1972 and several locomotives have carried more than one number in this scheme.

Locomotives in this section are listed in 1957 number order within each class. A number of diesel shunting locomotives were withdrawn prior to the classification system being introduced and are listed at the end of this section. Experimental and civil engineers main line locomotives are listed in sections 2.4 and 2.5.

Classification System

It was not until the British Railways organisation was set up that some semblance of order was introduced. This broadly took the following form:

Type	Engine Horsepower	Number Range
1	800–1000	D 8000–D 8999
2	1001–1499	D 5000–D 6499
3	1500–1999	D 6500–D 7999
4	2000–2999	D 1–D 1999
5	3000 +	D 9000–D 9499
Shunting	150/300	D 2000–D 2999
Shunting	350/400	D 3000–D 4999
Shunting/trip	650	D 9500–D 9999
AC Electric		E 1000–E 4999
DC Electric		E 5000–E 6999

Note: Locomotives which have recently been preserved, but have not yet moved from the National Rail network are shown with the location left blank.

CLASS 01   0-4-0

Built: 1956 by Andrew Barclay, Kilmarnock. 4 built.
Engine: Gardner 6L3 of 114 kW (153 h.p.) at 1200 r.p.m.
Transmission: Mechanical. Wilson SE4 epicyclic
Maximum Tractive Effort: 56.8 kN (12750 lbf).   **Weight:** 25.5 tonnes.
Wheel Diameter: 965 mm.   **Maximum Speed:** 14 m.p.h.
No train brakes.

11503–D 2953	Peak Railway	AB 395/1956
11506–D 2956	East Lancashire Railway	AB 398/1956

CLASS 02   0-4-0

Built: 1960–61 By Yorkshire Engine Company, Sheffield. 20 built.
Engine: Rolls Royce C6NFL of 127 kW (170 h.p.) at 1800 r.p.m.
Transmission: Hydraulic. Rolls Royce CF 10000.
Maximum Tractive Effort: 66.8 kN (15000 lbf).
Continuous Tractive Effort: 61 kN (13700 lbf) at 1.4 m.p.h.
Weight: 28.6 tonnes.
Wheel Diameter: 1067 mm.   **Maximum Speed:** 30 m.p.h.

BR	Present		
D 2853–02003	"PETER"	Appleby-Frodingham RPS	YE 2812/1960
D 2854		Peak Railway	YE 2813/1960
D 2858	Unnumbered	Midland Railway-Butterley	YE 2817/1960
D 2860		National Railway Museum	YE 2843/1961
D 2866	Unnumbered	Peak Railway	YE 2849/1961
D 2867	"DIANNE"	Battlefield Railway	YE 2850/1961
D 2868	"SAM"	Barrow Hill Roundhouse	YE 2851/1961

CLASS 03   0-6-0

Built: 1957–62 at Doncaster and Swindon. 230 built.
Engine: Gardner 8L3 of 152 kW (204 h.p.) at 1200 r.p.m. Replaced with VM V12 (300–350 h.p.) on D 2128 and D 2134.
Transmission: Mechanical. Wilson CA5 epicyclic gearbox.
Maximum Tractive Effort: 68 kN (15300 lbf).   **Weight:** 31 tonnes.
Wheel Diameter: 1092 mm.   **Maximum Speed:** 28 m.p.h.
x Dual-braked   a Air brakes only

§ modified with cut down cab for working on Burry Port & Gwendraeth Valley Line.

11205–D 2018–03018		Mangapps Farm, Burnham-on-Crouch	Swindon 1958
11207–D 2020–03020		Lavender Line, Isfield	Swindon 1958
11209–D 2022–03022		Swindon & Cricklade Railway	Swindon 1958
11210–D 2023		Kent & East Sussex Railway	Swindon 1958
11211–D 2024	No. 4	Kent & East Sussex Railway	Swindon 1958
D 2027–03027		Peak Railway	Swindon 1958
D 2033	Profilatinave 2	Siderurgica, Montirone, Brescia, Italy	Swindon 1958
D 2036	Profilatinave 1	Siderurgica, Montirone, Brescia, Italy	Swindon 1959
D 2037–03037		Peak Railway	Swindon 1959
D 2041		Colne Valley Railway	Swindon 1959
D 2046	Unnumbered	Rye Farm, Wishaw	Doncaster 1958
D 2051		North Norfolk Railway	Doncaster 1959
D 2059–03059x		Isle of Wight Steam Railway	Doncaster 1959
D 2062–03062		East Lancashire Railway	Doncaster 1959
D 2063–03063x		North Norfolk Railway	Doncaster 1959
D 2066–03066x		Barrow Hill Roundhouse	Doncaster 1959
D 2069–03069		Gloucestershire-Warwickshire Railway	Doncaster 1959
D 2072–03072		Lakeside & Haverthwaite Railway	Doncaster 1959
D 2073–03073x		The Railway Age, Crewe	Doncaster 1959
D 2078–03078x		Stephenson Railway Museum	Doncaster 1959
D 2079–03079		Derwent Valley Light Railway	Doncaster 1960
D 2081–03081		Mangapps Farm, Burnham-on-Crouch	Doncaster 1960

89

D 2084–03084x	"HELEN-LOUISE"	Ecclesbourne Valley Railway	Doncaster 1959
D 2089–03089x		Mangapps Farm, Burnham-on-Crouch	Doncaster 1960
D 2090–03090		Locomotion: The NRM at Shildon	Doncaster 1960
D 2094–03094x		Cambrian Railway Trust, Llynclys	Doncaster 1960
D 2099–03099		Peak Railway	Doncaster 1960
D 2113–03113		Peak Railway	Doncaster 1960
D 2117	LHR No. 8	Lakeside & Haverthwaite Railway	Swindon 1959
D 2118		Peak Railway	Swindon 1959
D 2119–03119§	"LINDA"	West Somerset Railway	Swindon 1959
D 2120–03120§		Fawley Hill Railway	Swindon 1959
D 2128–03128a		Dean Forest Railway	Swindon 1960
D 2133		West Somerset Railway	Swindon 1960
D 2134–03134a		Royal Deeside Railway	Swindon 1960
D 2138		Midland Railway-Butterley	Swindon 1960
D 2139	No. 1	Peak Railway	Swindon 1960
D 2141–03141§		Swansea Vale Railway	Swindon 1960
D 2144–03144§x	"WESTERN WAGGONER"	Wensleydale Railway	Swindon 1960
D 2145–03145§		Moreton Park Railway	Swindon 1960
D 2148		Ribble Steam Railway	Swindon 1960
D 2152–03152§	Unnumbered	Swindon & Cricklade Railway	Swindon 1960
D 2158–03158x	"MARGARET ANN"	Ecclesbourne Valley Railway	Swindon 1960
D 2162–03162x		Llangollen Railway	Swindon 1960
D 2170–03170x		Battlefield Railway	Swindon 1960
D 2178	2	Gwili Railway	Swindon 1962
D 2180–03180x		Battlefield Railway	Swindon 1962
D 2182		Gloucestershire-Warwickshire Railway	Swindon 1962
D 2184		Colne Valley Railway	Swindon 1962
D 2189–03189		Ribble Steam Railway	Swindon 1961
D 2192	"TITAN"	Paignton & Dartmouth Railway	Swindon 1961
D 2197–03197x		Lavender Line, Isfield	Swindon 1961
D 2199		Peak Railway	Swindon 1961
Dept 92–D 2371–03371x		Rowden Mill Station, Herefordshire	Swindon 1958
D 2399–03399x		Mangapps Farm, Burnham-on-Crouch	Doncaster 1961

Note: 03144 is on loan from MoD 275 Squadron RLC (V).

CLASS 04   0-6-0

Built: 1952–62. Drewry design built by Vulcan Foundry & Robert Stephenson & Hawthorn. 140 built.
Engine: Gardner 8L3 of 152 kW (204 h.p.) at 1200 r.p.m.
Transmission: Mechanical. Wilson CA5 epicyclic gearbox
Maximum Tractive Effort: 69.7 kN (15650 lbf). **Weight:** 32 tonnes.
Wheel Diameter: 1067 mm.   **Maximum Speed:** 28 m.p.h.

BR	Present		
11103–D 2203		Embsay & Bolton Abbey Railway	DC/VF 2400/D145/1952
11106–D 2205		West Somerset Railway	DC/VF 2486/D212/1953
11108–D 2207		North Yorkshire Moors Railway	DC/VF 2482/D208/1953
11135–D 2229	No. 5	Peak Railway	DC/VF 2552/D278/1955
11215–D 2245		Battlefield Railway	DC/RSH 2577/7864/1956
11216–D 2246		South Devon Railway	DC/RSH 2578/7865/1956
D 2271		West Somerset Railway	DC/RSH 2615/7913/1958
D 2272	"ALFIE"	Peak Railway	DC/RSH 2616/7914/1958
D 2279		East Anglian Railway Museum	DC/RSH 2656/8097/1960
D 2280	No. 2	North Norfolk Railway	DC/RSH 2657/8098/1960
D 2284		Peak Railway	DC/RSH 2661/8102/1960
D 2289		Acciaierie, Lonato, Italy	DC/RSH 2669/8122/1960
D 2298		Buckinghamshire Railway Centre	DC/RSH 2679/8157/1960
D 2302		Barrow Hill Roundhouse	DC/RSH 2683/8161/1960
D 2310		Battlefield Railway	DC/RSH 2691/8169/1960
D 2324	"Judith"	Barrow Hill Roundhouse	DC/RSH 2705/8183/1961
D 2325		Mangapps Farm, Burnham-on-Crouch	DC/RSH 2706/8184/1961
D 2334	NCB 33 NO. 4	Churnet Valley Railway	DC/RSH 2715/8193/1961
D 2337	"DOROTHY"	Peak Railway	DC/RSH 2718/8196/1961

CLASS 05   0-6-0

Built: 1955–61 by Hunslet Engine Company, Leeds. 69 built.
Engine: Gardner 8L3 of 152 kW (204 h.p.) at 1200 r.p.m.
Transmission: Mechanical. Hunslet gearbox.
Maximum Tractive Effort: 64.6 kN (14500 lbf).   **Weight:** 31 tonnes.
Wheel Diameter: 1121 (1016*) mm.   **Maximum Speed:** 18 m.p.h.

BR	Present		
11140–D 2554–05001–97803*		Isle of Wight Steam Railway	HE 4870/1956
D 2578	2 "CIDER QUEEN"	Moreton Park Railway	HE 5460/1958 rebuilt HE 6999/1968
D 2587		Peak Railway	HE 5636/1959 rebuilt HE 7180/1969
D 2595		Ribble Steam Railway	HE 5644/1960 rebuilt HE 7179/1969

CLASS 06   0-4-0

Built: 1958–60 by Andrew Barclay, Kilmarnock. 35 built.
Engine: Gardner 8L3 of 152 kW (204 h.p.) at 1200 r.p.m.
Transmission: Mechanical. Wilson CA5 epicyclic gearbox.
Maximum Tractive Effort: 88 kN (19800 lbf).   **Weight:** 37 tonnes.
Wheel Diameter: 1092 mm.   **Maximum Speed:** 23 m.p.h.

D 2420–06003–97804   Barrow Hill Roundhouse   AB 435/1959

CLASS 07   0-6-0

Built: 1962 by Ruston & Hornsby, Lincoln. 14 built.
Engine: Paxman 6RPHL Mk III of 205 kW (275 h.p.) at 1360 r.p.m.
Transmission: Electric. One AEI RTB 6652 traction motor.
Maximum Tractive Effort: 126 kN (28240 lbf).
Continuous Tractive Effort: 71 kN (15950 lbf) at 4.38 m.p.h.
Power at Rail: 142 kW (190 h.p.).   **Weight:** 43.6 tonnes.
Wheel Diameter: 1067 mm.   **Maximum Speed:** 20 m.p.h.

x-Dual-braked.

BR	Present		
D 2989–07005x	Unnumbered	Battlefield Railway	RH 480690/1962
D 2994–07010		Avon Valley Railway	RH 480695/1962
D 2995–07011x	"CLEVELAND"	St. Leonards Railway Engineering	RH 480696/1962
D 2996–07012		Barrow Hill Roundhouse	RH 480697/1962
D 2997–07013x		Peak Railway	RH 480698/1962

CLASS 08   0-6-0

Built: 1952–62. Built at Derby, Darlington, Crewe, Horwich & Doncaster. 996 built.
Engine: English Electric 6KT of 298 kW (400 h.p.) at 680 r.p.m.
Transmission: Electric. Two EE 506 axle-hung traction motors.
Maximum Tractive Effort: 156 kN (35000 lbf).
Continuous Tractive Effort: 49.4 kN (11100 lbf) at 8.8 m.p.h.
Power at Rail: 194 kW (260 h.p.).   **Weight:** 50 tonnes.
Wheel Diameter: 1372 mm.   **Maximum Speed:** 15 m.p.h.

x-Dual-braked.   a-Air-braked only.

BR	Present			
13000–D 3000			Barrow Hill Roundhouse	Derby 1952
13002–D 3002			Plym Valley Railway	Derby 1952
13014–D 3014		"SAMSON"	Paignton & Dartmouth Railway	Derby 1952
13018–D 3018–08011			Chinnor & Princes Risborough Railway	Derby 1953
13019–D 3019			Cambrian Railway Trust, Llynclys	Derby 1953
13022–D 3022–08015			Severn Valley Railway	Derby 1953
13023–D 3023–08016		"Geoff L. Wright"	Bluebell Railway	Derby 1953
13029–D 3029–08021			Tyseley Locomotive Works, Birmingham	Derby 1953
13030–D 3030–08022		"LION"	Cholsey & Wallingford Railway	Derby 1953
13059–D 3059–08046		"Brechin City"	Caledonian Railway	Derby 1954
13074–D 3074–08060		"UNICORN"	Cholsey & Wallingford Railway	Darlington 1953
13079–D 3079–08064			National Railway Museum	Darlington 1954
13101–D 3101			Great Central Railway	Derby 1955
13167–D 3167–08102			Lincolnshire Wolds Railway	Derby 1955
13174–D 3174–08108		"Dover Castle"	Kent & East Sussex Railway	Derby 1955
13180–D 3180–08114			Nottingham Heritage Centre, Ruddington	Derby 1955
13190–D 3190–08123			Cholsey & Wallingford Railway	Derby 1955
13201–D 3201–08133			Severn Valley Railway	Derby 1955
13232–D 3232–08164			East Lancashire Railway	Darlington 1956
13236–D 3236–08168			Battlefield Railway	Darlington 1956
13255–D 3255			Brighton Railway Museum	Derby 1956
13261–D 3261			Swindon & Cricklade Railway	Derby 1956
13265–D 3265–08195		"MARK"	Llangollen Railway	Derby 1956
13290–D 3290–08220		Unnumbered	Wrenbury Business Park, Cheshire	Derby 1956
13308–D 3308–08238		"Charlie"	Dean Forest Railway	Darlington 1956
13336–D 3336–08266			Keighley & Worth Valley Railway	Darlington 1957
D 3358–08288		Unnumbered	Mid Hants Railway	Derby 1957
D 3420–08350			Midland Railway-Butterley	Crewe 1957
D 3429–08359			Telford Steam Railway	Crewe 1958
D 3462–08377			West Somerset Railway	Darlington 1957
D 3508–08388			Reliance Industrial Estate, Manchester	Derby 1958
D 3551–08436 x			Swanage Railway	Derby 1958
D 3558–08443			Bo'ness & Kinneil Railway	Derby 1958
D 3559–08444			Bodmin Steam Railway	Derby 1958
D 3560–08445 a			East Lancashire Railway	Derby 1958
D 3586–08471			Severn Valley Railway	Crewe 1958
D 3588–08473			Dean Forest Railway	Crewe 1958
D 3591–08476			Swanage Railway	Crewe 1958
D 3594–08479			East Lancashire Railway	Horwich 1958
D 3596–08481 x			Barry Island Railway	Horwich 1958
D 3605–08490			Strathspey Railway	Horwich 1958
D 3673–08511 a			Barry Island Railway	Darlington 1958
D 3723–08556			North Yorkshire Moors Railway	Darlington 1959
D 3757–08590 x			Midland Railway-Butterley	Crewe 1959
D 3771–08604 x		604 "PHANTOM"	Didcot Railway Centre	Derby 1959
D 3795–08628 x			Bryn Engineering, Blackrod	Derby 1959
D 3798–08631 x		"EAGLE"	Mid Norfolk Railway	Derby 1959
D 3802–08635 x			Severn Valley Railway	Derby 1959
D 3867–08700 a			East Lancashire Railway	Horwich 1960
D 3902–08734 x			Dean Forest Railway	Crewe 1960
D 3935–08767 x			North Norfolk Railway	Horwich 1961
D 3937–08769		"GLADYS"	Dean Forest Railway	Derby 1960
D 3940–08772 x		"CAMULODUNUM"	North Norfolk Railway	Derby 1960
D 3941–08773 x			Embsay & Bolton Abbey Railway	Derby 1960
D 3991–08823 a			Churnet Valley Railway	Derby 1960
D 3993–08825 a			Battlefield Railway	Derby 1960
D 4018–08850 x			North Yorkshire Moors Railway	Horwich 1961
D 4141–08911 x		"MATEY"	National Railway Museum	Horwich 1962
D 4157–08927 x			Gloucestershire-Warwickshire Railway	Horwich 1962
D 4174–08944 x			East Lancashire Railway	Darlington 1962

Note: 08473 is being dismantled and used as a source of spares.

In addition Locomotives of a similar design were constructed by English Electric for Netherlands Railways (NS). Several of these have now returned to Great Britain and those in preservation are listed here:

NS CLASS 600   0-6-0

Built: 1955–1956. Built by English Electric. 65 built.
Engine: English Electric 6KT of 298 kW 400 h.p.) at 680 r.p.m.
Transmission: Electric: Two EE506 axle-hung traction motors.
Power at Rail: 191 kW (255 h.p.)
Maximum Tractive Effort: 143 kN (32000 lbf).   **Weight:** 47 tonnes.
Continuous Tractive Effort: 54 kN at 7.2 m.p.h. **Wheel Diameter:** 1232 mm.
Maximum Speed: 18.75 m.p.h.   Air-braked.

601–671	Ribble Steam Railway	EE/DK 2098 1955
663	Locomotion: The NRM at Shildon	EE/VF 2160/D350 1956

CLASS 09   0-6-0

Built: 1959–62. Built at Darlington & Horwich. 26 built.
Engine: English Electric 6KT of 298 kW (400 h.p.) at 680 r.p.m.
Transmission: Electric: Two English Electric EE506 axle-hung traction motors.
Power at Rail: 201 kW (269 h.p.).
Maximum Tractive Effort: 111 kN (25000 lbf).
Continuous Tractive Effort: 39 kN (8800 lbf) at 11.6 m.p.h.
Weight: 50 tonnes.   **Wheel Diameter:** 1372 mm.
Maximum Speed: 27 m.p.h.   Dual-braked.

D 3666–09002	South Devon Railway	Darlington 1959
D 3668–09004	Spa Valley Railway	Darlington 1959
D 4113–09025	East Kent Light Railway	Horwich 1962

CLASS 10   0-6-0

Built: 1955–62. Built at Darlington & Doncaster. 146 built.
Engine: Lister Blackstone ER6T of 261 kW (350 h.p.) at 750 r.p.m.
Transmission: Electric. Two GEC WT 821 axle-hung traction motors.
Power at Rail: 198 kW (265 h.p.).
Maximum Tractive Effort: 156 kN (35000 lbf).   **Weight:** 47 tonnes.
Continuous Tractive Effort: 53.4 kN (12000 lbf) at 8.2 m.p.h.
Wheel Diameter: 1372 mm.   **Maximum Speed:** 20 m.p.h.

D 3452		Bodmin Steam Railway	Darlington 1957
D 3476		Colne Valley Railway	Darlington 1957
D 3489	"COLONEL TOMLINE"	Spa Valley Railway	Darlington 1958
D 4067	"MARGARET ETHEL – THOMAS ALFRED NAYLOR"		
		Great Central Railway	Darlington 1961
D 4092	"CHRISTINE"	Barrow Hill Roundhouse	Darlington 1962

CLASS 14 — 0-6-0

Built: 1964–65 at Swindon. 56 built.
Engine: Paxman Ventura 6YJXL of 485 kW (650 h.p.) at 1500 r.p.m.
Transmission: Hydraulic. Voith L217u
Maximum Tractive Effort: 135 kN (30910 lbf).
Continuous Tractive Effort: 109 kN (26690 lbf) at 5.6 m.p.h.
Weight: 51 tonnes. **Wheel Diameter**: 1219 mm.
Maximum Speed: 40 m.p.h. **Train Heating**: None.

x Dual-braked

BR	Present		
D 9500	9312/92	Barrow Hill Roundhouse	Swindon 1964
D 9502	9312/97	Peak Railway	Swindon 1964
D 9513	N.C.B. 38	Embsay & Bolton Abbey Railway	Swindon 1964
D 9516 x		Nene Valley Railway	Swindon 1964
D 9518	N.C.B. 7	Nene Valley Railway	Swindon 1964
D 9520		Nene Valley Railway	Swindon 1964
D 9521		Barry Island Railway	Swindon 1964
D 9523 x		Nene Valley Railway	Swindon 1964
D 9525		Peak Railway	Swindon 1965
D 9526		West Somerset Railway	Swindon 1965
D 9531		East Lancashire Railway	Swindon 1965
D 9537		Rippingale Station, Lincolnshire	Swindon 1965
D 9539		Ribble Steam Railway	Swindon 1965
D 9551		Royal Deeside Railway	Swindon 1965
D 9553	54	Gloucestershire-Warwickshire Railway	Swindon 1965
D 9555		Dean Forest Railway	Swindon 1965

Note: D 9555 is on loan from the Rutland Railway Museum.

CLASS 15 — Bo-Bo

Built: 1957–59 by BTH/Clayton Equipment Company. 44 built.
Engine: Paxman 16YHXL of 597 kW (800 h.p.) at 1250 r.p.m.
Transmission: Electric. Four BTH 137AZ axle-hung traction motors.
Power at Rail:
Maximum Tractive Effort: 178 kN (40000 lbf).
Continuous Tractive Effort: 88 kN (19700 lbf) at 11.3 m.p.h.
Weight: 69 tonnes. **Wheel Diameter**: 1003 mm.
Maximum Speed: 60 m.p.h. **Train Heating**: None.

D 8233–ADB968001	East Lancashire Railway	BTH 1131/1960

CLASS 17 — Bo-Bo

Built: 1962–65 by Clayton Equipment Company. 117 built.
Engine: Two Paxman 67HXL of 336 kW (450 h.p.) at 1500 r.p.m.
Transmission: Electric. Four GEC WT421 axle-hung traction motors.
Power at Rail: 461 kW (618 h.p.).
Maximum Tractive Effort: 178 kN (40000 lbf).
Continuous Tractive Effort 80 kN (18000 lbf) at 12.8 m.p.h.
Weight: 69 tonnes. **Wheel Diameter**: 1003 mm.
Maximum Speed: 60 m.p.h. **Train Heating**: None.

D 8568	Chinnor & Princes Risborough Railway	CE 4365U/69 1964

CLASS 20 — Bo-Bo

Built: 1957–68 by English Electric at Vulcan Foundry, Newton-le-Willows or Robert Stephenson & Hawthorn, Darlington. 228 built.
Engine: English Electric 8SVT of 746 kW (1000 h.p.) at 850 r.p.m.
Transmission: Electric. Four EE 526/5D axle-hung traction motors.
Power at Rail: 574 kW (770 h.p.).
Maximum Tractive Effort: 187 kN (42000 lbf).
Continuous Tractive Effort: 111 kN (25000 lbf) at 11 m.p.h.
Weight: 74 tonnes. **Wheel Diameter:** 1092 mm.
Maximum Speed: 75 m.p.h. **Train Heating:** None.

Dual braked except D 8000.

D 8000–20050		National Railway Museum	EE/VF 2347/D375 1957
D 8001–20001		Midland Railway-Butterley	EE/VF 2348/D376 1957
D 8007–20007		Churnet Valley Railway	EE/VF 2354/D382 1957
D 8020–20020		Bo'ness & Kinneil Railway	EE/RSH 2742/8052 1959
D 8031–20031		Keighley & Worth Valley Railway	EE/RSH 2753/8063 1960
D 8035–20035	2001	Rye Farm, Wishaw	EE/VF 2757/D482 1959
D 8048–20048		Nottingham Heritage Centre	EE/VF 2770/D495 1959
D 8059–20059		Tyseley Locomotive Works	EE/RSH 2965/8217 1961
D 8063–20063	2002	Rye Farm, Wishaw	EE/RSH 2969/8221 1961
D 8069–20069		Mid Norfolk Railway	EE/RSH 2975/8227 1961
D 8087–20087		East Lancashire Railway	EE/RSH 2993/8245 1961
D 8098–20098		Great Central Railway	EE/RSH 3004/8256 1961
D 8107–20107	H010	Weardale Railway	EE/RSH 3013/8265 1961
D 8110–20110		South Devon Railway	EE/RSH 3016/8268 1962
D 8118–20118	Saltburn-by-the-Sea	South Devon Railway	EE/RSH 3024/8276 1962
D 8128–20228	2004	Barry Island Railway	EE/VF 3599/D998 1966
D 8137–20137	Murray B. Hofmeyr	Gloucestershire-Warwickshire Rly	EE/VF 3608/D1007 1966
D 8139–20139	2003	Rye Farm, Wishaw	EE/VF 3610/D1009 1966
D 8142–20142		Llangollen Railway	EE/VF 3613/D1013 1966
D 8154–20154		Nottingham Heritage Centre	EE/VF 3625/D1024 1966
D 8166–20166	"RIVER FOWEY"	Bodmin Steam Railway	EE/VF 3637/D1036 1966
D 8169–20169		Stainmore Railway, Kirkby Stephen East	EE/VF 3640/D1039 1966
D 8177–20177		Tyseley Locomotive Works	EE/VF 3648/D1047 1966
D 8188–20188		Severn Valley Railway	EE/VF 3669/D1064 1967
D 8189–20189		Embsay & Bolton Abbey Railway	EE/VF 3670/D1065 1967
D 8197–20197		Bodmin Steam Railway	EE/VF 3678/D1073 1967
D 8305–20205		Midland Railway-Butterley	EE/VF 3686/D1081 1967
D 8314–20214		Lakeside & Haverthwaite Railway	EE/VF 3695/D1090 1967
D 8327–20227		Midland Railway-Butterley	EE/VF 3685/D1080 1968

CLASS 24 — Bo-Bo

Built: 1958–61 at Derby, Crewe & Darlington. 151 built.
Engine: Sulzer 6LDA28A of 870 kW (1160 h.p.) at 750 r.p.m.
Transmission: Electric. Four BTH 137BY axle-hung traction motors.
Power at Rail: 629 kW (843 h.p.).
Maximum Tractive Effort: 178 kN (40000 lbf).
Continuous Tractive Effort: 95 kN (21300 lbf) at 4.38 m.p.h.
Weight: 78 (81*) tonnes. **Wheel Diameter:** 1143 mm.
Maximum Speed: 75 m.p.h. **Train Heating:** Steam.

D 5032–24032*		North Yorkshire Moors Railway	Crewe 1959
D 5054–24054–ADB 968008	"Phil Southern"	East Lancashire Railway	Crewe 1959
D 5061–24061–RDB 968007-97201		North Yorkshire Moors Railway	Crewe 1960
D 5081–24081		Gloucestershire-Warwickshire Railway	Crewe 1960

CLASS 25   Bo-Bo

Built: 1961–67. Built at Darlington, Derby & Beyer Peacock. 327 built.
Engine: Sulzer 6LDA28B of 930 kW (1250 h.p.) at 750 r.p.m.
Transmission: Electric. Four AEI 253AY axle-hung traction motors.
Power at Rail: 708 kW (949 h.p.).
Maximum Tractive Effort: 200 kN (45000 lbf).
Continuous Tractive Effort: 93 kN (20800 lbf) at 17.1 m.p.h.
Weight: 72–76 tonnes.   **Wheel Diameter:** 1143 mm.
Maximum Speed: 90 m.p.h.

Class 25/1. Dual-braked except D 5217. Train heating: Steam.

D 5185–25035		Great Central Railway	Darlington 1963
D 5207–25057		North Norfolk Railway	Derby 1963
D 5209–25059		Keighley & Worth Valley Railway	Derby 1963
D 5217–25067		Barrow Hill Roundhouse	Derby 1963
D 5222–25072		Caledonian Railway	Derby 1963

Class 25/2. Dual-braked (except D 5233). Train heating: Steam: D 5233, D 7585/94. None: D7523/35/41.

D 5233–25083		Caledonian Railway	Derby 1963
D 7523–25173	"John F Kennedy"	West Somerset Railway	Derby 1965
D 7535–25185	"HERCULES"	Paignton & Dartmouth Railway	Derby 1965
D 7541–25191	"THE DIANA"	North Yorkshire Moors Railway	Derby 1965
D 7585–25235		Bo'ness & Kinneil Railway	Darlington 1964
D 7594–25244		Kent & East Sussex Railway	Darlington 1964

Class 25/3. Dual-braked. Train heating: None.

D 7612–25262–25901		South Devon Railway	Derby 1966
D 7615–25265		Great Central Railway	Derby 1966
D 7628–25278	"SYBILLA"	North Yorkshire Moors Railway	BP 8038/1965
D 7629–25279		Nottingham Heritage Centre	BP 8039/1965
D 7633–25283–25904		Dean Forest Railway	BP 8043/1965
D 7659–25309–25909		Bo'ness & Kinneil Railway	BP 8069/1966
D 7663–25313	"Chirk Castle"	Llangollen Railway	Derby 1966
D 7671–25321		Midland Railway-Butterley	Derby 1967
D 7672–25322–25912	"TAMWORTH CASTLE"	Churnet Valley Railway	Derby 1967

Note: 25035 is on loan from the Northampton & Lamport Railway.

CLASS 26   Bo-Bo

Built: 1958–59 by the Birmingham Railway Carriage & Wagon Company. 47 built.
Engine: Sulzer 6LDA28B of 870 kW (1160 h.p.) at 750 r.p.m.
Transmission: Electric. Four Crompton-Parkinson C171A1 (C171D3§) axle-hung traction motors.
Power at Rail: 671 kW (900 h.p.).
Maximum Tractive Effort: 187 kN (42000 lbf).
Continuous Tractive Effort: 133 kN (30000 lbf) at 14 m.p.h.
Weight: 72–75 tonnes.   **Wheel Diameter:** 1092 mm.
Maximum Speed: 75 m.p.h.   Dual-Braked.
Train Heating: Built with steam. Removed from D 5300/01/02/04 in 1967.

Class 26/0.

D 5300–26007	Barrow Hill Roundhouse	BRCW DEL/45/1958
D 5301–26001	Lakeside & Haverthwaite Railway	BRCW DEL/46/1958
D 5302–26002	Strathspey Railway	BRCW DEL/47/1958
D 5304–26004	Bo'ness & Kinneil Railway	BRCW DEL/49/1958
D 5310–26010	Bo'ness & Kinneil Railway	BRCW DEL/55/1959
D 5311–26011	Barrow Hill Roundhouse	BRCW DEL/56/1959
D 5314–26014	Caledonian Railway	BRCW DEL/59/1959

Class 26/1.§

D 5324–26024	Bo'ness & Kinneil Railway	BRCW DEL/69/1959
D 5325–26025	Strathspey Railway	BRCW DEL/70/1959
D 5335–26035	Caledonian Railway	BRCW DEL/80/1959
D 5338–26038	Pullman Design & Fabrication, Cardiff	BRCW DEL/83/1959
D 5340–26040	Kingdom of Fife Railway	BRCW DEL/85/1959
D 5343–26043	Gloucestershire-Warwickshire Railway	BRCW DEL/88/1959

Note: 26040 is currently receiving attention at Barclay Brown Yard, Methil.

CLASS 27   Bo-Bo

Built: 1961–62 by the Birmingham Railway Carriage & Wagon Company, Birmingham. 69 built.
Engine: Sulzer 6LDA28B of 930 kW (1250 h.p.) at 750 r.p.m.
Transmission: Electric. Four GEC WT459 axle-hung traction motors.
Power at Rail: 696 kW (933 h.p.).
Maximum Tractive Effort: 178 kN (40000 lbf).
Continuous Tractive Effort: 111 kN (25000 lbf) at 14 m.p.h.
Weight: 72–75 tonnes.   **Wheel Diameter:** 1092 mm.
Maximum Speed: 90 m.p.h.
Train Heating: Built with steam (except D 5370 – no provision). Replaced with electric on D 5386 and D 5410 but subsequently removed.

Dual-braked except D 5353.

D 5347–27001	Bo'ness & Kinneil Railway	BRCW DEL/190/1961
D 5351–27005	Bo'ness & Kinneil Railway	BRCW DEL/194/1961
D 5353–27007	Mid Hants Railway	BRCW DEL/196/1961
D 5370–27024–ADB 968028	Caledonian Railway	BRCW DEL/213/1962
D 5386–27103–27212–27066	Dean Forest Railway	BRCW DEL/229/1962
D 5394–27106–27050	Bo'ness & Kinneil Railway	BRCW DEL/237/1962
D 5401–27112–27056	Northampton & Lamport Railway	BRCW DEL/244/1962
D 5410–27123–27205–27059	Severn Valley Railway	BRCW DEL/253/1962

Note: 27050 is on loan from the Strathspey Railway.

CLASS 28   METROVICK   Co-Bo

Built: 1958–59 by Metropolitan Vickers, Manchester. 20 built.
Engine: Crossley HSTVee 8 of 896 kW (1200 h.p.) at 625 r.p.m.
Transmission: Electric. Five MV 137BZ axle-hung traction motors.
Power at Rail: 671 kW (900 h.p.).
Maximum Tractive Effort: 223 kN (50000 lbf).   **Weight:** 99 tonnes.
Continuous Tractive Effort: 111 kN (25000 lbf) at 13.5 m.p.h.   **Wheel Diameter:** 1003 mm.
Maximum Speed: 75 m.p.h.   **Train Heating:** Steam.

| D 5705–S 15705–TDB 968006 | East Lancashire Railway | MV 1958 |

CLASS 31　　　　　　　　　　　　　　　　　　　　　　　　　　　　　　　　A1A-A1A

Built: 1957–62 by Brush Electrical Engineering Company, Loughborough. 263 built.
Engine: Built with Mirrlees JVS12T of 930 kW (1250 h.p.). Re-engined 1964–9 with English Electric 12SVT of 1100 kW (1470 h.p.) at 850 r.p.m.
Transmission: Electric. Four Brush TM73-68 axle-hung traction motors.
Power at Rail: 872 kW (1170 h.p.).
Maximum Tractive Effort: 190 kN (42800 lbf).
Continuous Tractive Effort: 99 kN (22250 lbf) at 19.7 m.p.h.
Weight: 110 tonnes.　　　　　　　　　　　　**Wheel Diameter:** 1092 mm.
Maximum Speed: 90 (80§) m.p.h. D 5518, D 5522, D 5526, D 5531 and D 5533 were originally 80 m.p.h. but have been regeared for 90 m.p.h.
Train Heating: Built with steam heating. Electric heating fitted and steam heating removed on D 5522, D5533, D 5547, D 5557, D 5600, D 5641, D 5669, D 5695, D 5814, D 5823, D 5830 and D 5844.
Note: D 5500, D 5518, D 5522, D 5526, D5547 and D 5562 were built without roof-mounted headcode boxes. They were subsequently fitted to D 5518.

Class 31/0. Electromagnetic Control.

D 5500–31018§	National Railway Museum	BE 71/1957

Class 31/1. Electro-pneumatic Control. Dual braked.

D 5518–31101 "Brush Veteran"	Battlefield Railway	BE 89/1958
D 5522–31418	Midland Railway-Butterley	BE 121/1959
D 5526–31108	South Devon Railway	BE 125/1959
D 5531–31113	Chinnor & Princes Risborough Railway	BE 130/1959
D 5533–31115–31466	Dean Forest Railway	BE 132/1959
D 5537–31119	Embsay & Bolton Abbey Railway	BE 136/1959
D 5547–31129–31461	Battlefield Railway	BE 146/1959
D 5548–31130 "Calder Hall Power Station"	Battlefield Railway	BE 147/1959
D 5557–31139–31438–31538	Mid Norfolk Railway	BE 156/1959
D 5562–31144	Reliance Industrial Estate, Manchester	BE 161/1959
D 5580–31162	North Norfolk Railway	BE 180/1960
D 5581–31163	Chinnor & Princes Risborough Railway	BE 181/1960
D 5584–31166	Wensleydale Railway	BE 184/1960
D 5600–31179–31435 "Newton Heath TMD"	East Lancashire Railway	BE 200/1960
D 5611–31188	Wensleydale Railway	BE 211/1960
D 5627–31203	Chasewater Railway	BE 227/1960
D 5630–31206	Rushden Station Museum	BE 230/1960
D 5631–31207	North Norfolk Railway	BE 231/1960
D 5634–31210	Dean Forest Railway	BE 234/1960
D 5641–31216–31467	East Lancashire Railway	BE 241/1961
D 5662–31235	Mid Norfolk Railway	BE 262/1960
D 5669–31410 Granada Telethon	Stainmore Railway, Kirby Stephen East	BE 269/1960
D 5683–31255	Colne Valley Railway	BE 284/1961
D 5695–31265–31430–31530 Sister Dora	Mid Norfolk Railway	BE 296/1961
D 5800–31270	Peak Railway	BE 301/1961
D 5801–31271 "Stratford 1840–2001"	Nene Valley Railway	BE 302/1961
D 5814–31414–31514	Ecclesbourne Valley Railway	BE 315/1961
D 5821–31289 "PHOENIX"	Northampton & Lamport Railway	BE 322/1961
D 5823–31291–31456–31556	East Lancashire Railway	BE 324/1961
D 5830–31297–31463–31563	Great Central Railway	BE 366/1962
D 5844–31310–31422–31522	Tyseley Locomotive Works, Birmingham	BE 380/1962
D 5862–31327 Phillips-Imperial	Strathspey Railway	BE 398/1962

Note: 31108, 31162 and 31271 are on loan from the Midland Railway-Butterley.

CLASS 33   Bo-Bo

Built: 1961–62 by the Birmingham Railway Carriage & Wagon Company, Birmingham. 98 built.
Engine: Sulzer 8LDA28A of 1160 kW (1550 h.p.) at 750 r.p.m.
Transmission: Electric. Four Crompton-Parkinson C171C2 axle-hung traction motors.
Power at Rail: 906 kW (1215 h.p.).
Maximum Tractive Effort: 200 kN (45000 lbf).
Continuous Tractive Effort: 116 kN (26000 lbf) at 17.5 m.p.h.
Weight: 78 tonnes.   **Wheel Diameter:** 1092 mm.
Maximum Speed: 85 m.p.h.   **Train Heating:** Electric.
Dual-braked.

§ fitted for push-pull operation (Class 33/1).
* built to former loading gauge of the Tonbridge–Battle line (Class 33/2).

D 6501–33002	"Sea King"	South Devon Railway	BRCW DEL/93/1960
D 6508–33008	Eastleigh	Battlefield Railway	BRCW DEL/100/1960
D 6513–33102§		Churnet Valley Railway	BRCW DEL/105/1960
D 6515–33012	"Stan Symes"	Swanage Railway	BRCW DEL/107/1960
D 6521–33108§	VAMPIRE	Barrow Hill Roundhouse	BRCW DEL/113/1960
D 6525–33109§	CAPTAIN BILL SMITH RNR	LNWR, Crewe Carriage Shed	BRCW DEL/117/1960
D 6527–33110§		Bodmin Steam Railway	BRCW DEL/119/1960
D 6528–33111§		Barrow Hill Roundhouse	BRCW DEL/120/1960
D 6530–33018		Midland Railway-Butterley	BRCW DEL/122/1960
D 6534–33019	"Griffon"	Battlefield Railway	BRCW DEL/126/1960
D 6535–33116§	Hertfordshire Railtours	Great Central Railway (N)	BRCW DEL/127/1960
D 6536–33117§		East Lancashire Railway	BRCW DEL/128/1960
D 6539–33021	"Eastleigh"	Barrow Hill Roundhouse	BRCW DEL/131/1960
D 6552–33034		Swanage Railway	BRCW DEL/144/1961
D 6553–33035	Spitfire	Barrow Hill Roundhouse	BRCW DEL/145/1961
D 6564–33046	Merlin	Midland Railway-Butterley	BRCW DEL/156/1961
D 6566–33048		West Somerset Railway	BRCW DEL/170/1961
D 6570–33052	Ashford	Kent & East Sussex Railway	BRCW DEL/174/1961
D 6571–33053		Battlefield Railway	BRCW DEL/175/1961
D 6575–33057	Seagull	West Somerset Railway	BRCW DEL/179/1961
D 6583–33063	"R.J. Mitchell DESIGNER OF THE SPITFIRE"	Spa Valley Railway	BRCW DEL/187/1962
D 6585–33065	Sealion	Spa Valley Railway	BRCW DEL/189/1962
D 6586–33201*		Midland Railway-Butterley	BRCW DEL/157/1962
D 6593–33208*		Mid Hants Railway	BRCW DEL/164/1962

CLASS 35   HYMEK   B-B

Built: 1961–64. Beyer Peacock, Manchester. 101 built.
Engine: Maybach MD 870 of 1269 kW (1700 h.p.) at 1500 r.p.m.
Transmission: Hydraulic. Mekydro K184U.
Maximum Tractive Effort: 207 kN (46600 lbf).
Continuous Tractive Effort: 151 kN (33950 lbf) at 12.5 m.p.h.
Weight: 77 tonnes.   **Wheel Diameter:** 1143 mm.
Maximum Speed: 90 m.p.h.   **Train Heating:** Steam.

D 7017	West Somerset Railway	BP 7911/1962
D 7018	West Somerset Railway	BP 7912/1962
D 7029	Severn Valley Railway	BP 7923/1962
D 7076	East Lancashire Railway	BP 7980/1963

CLASS 37 Co-Co

Built: 1960–65 by English Electric Company at Vulcan Foundry, Newton-le-Willows or Robert Stephenson & Hawthorn, Darlington. 309 built.
Engine: English Electric 12CSVT of 1300 kW (1750 h.p.) at 850 r.p.m. (z re-engined with Mirrlees MB275T of 1340kw (1800hp)at 1000r.p.m.). (y re-engined with Ruston RK270T of 1340kW (1800 h.p. at 900 r.p.m.).
Transmission: Electric. Six English Electric 538/A.
Power at Rail: 932 kW (1250 h.p.).
Maximum Tractive Effort: 245 kN (55000 lbf).
Continuous Tractive Effort: 156 kN (35000 lbf) at 13.6 m.p.h.
Weight: 103–108 tonnes. (y z 120 tonnes)   **Wheel Diameter**: 1092 mm.
Maximum Speed: 80 m.p.h.
Train Heating: Steam (§ possibly built without heating, * built without heating, but steam later fitted to D 6963 and D 6964 and electric fitted to D 6987).
Dual-braked.

D 6700–37119–37350	"National Railway Museum"	North Yorkshire Moors Railway (N)	EE/VF 2863/D579/1960
D 6703–37003		Weardale Railway	EE/VF 2866/D582/1960
D 6709–37009–37340		Nottingham Heritage Centre	EE/VF 2872/D588/1961
D 6725–37025	"Inverness TMD"	Bo'ness & Kinneil Railway	EE/VF 2888/D604/1961
D 6728–37028–37505	British Steel Workington	Eden Valley Railway, Warcop	EE/VF 2891/D607/1961
D 6732–37032–37353		North Norfolk Railway	EE/VF 2895/D611/1962
D 6737–37037–37321	"Loch Treig"	South Devon Railway	EE/VF 2900/D616/1962
D 6775–37075		Churnet Valley Railway	EE/RSH 3067/8321/1962
D 6776–37076–37518		Nene Valley Railway	EE/RSH 3068/8322/1962
D 6797–37097		Caledonian Railway	EE/VF 3226/D751/1962
D 6799–37099–37324	Clydebridge	Gloucestershire-Warwickshire Rly	EE VF 3228/D753/1962
D 6816–37116	Sister Dora	Chinnor & Princes Risborough Railway	EE/VF 3245/D770/1963
D 6823–37123–37679§		Northampton & Lamport Railway	EE/RSH 3268/8383/1963
D 6836–37136–37905y§	Vulcan Enterprise	Dartmoor Railway, Okehampton	EE/VF 3281/D810/1963
D 6842–37142§		Bodmin Steam Railway	EE/VF 3317/D816/1963
D 6846–37146§		Stainmore Railway, Kirkby Stephen East	EE/VF 3321/D821/1963
D 6850–37150–37901z§	Mirrlees Pioneer	Llangollen Railway	EE/VF 3325/D824/1963
D 6852–37152–37310§	British Steel Ravenscraig	Peak Railway	EE/VF 3327/D826/1963
D 6869–37169–37674§	Saint Blaise Church 1445–1995	Stainmore Railway, Kirkby Stephen East	EE/VF 3347/D833/1963
D 6875–37175		Bo'ness & Kinneil Railway	EE/VF 3353/D839/1963
D 6888–37188	Jimmy Shand	Peak Railway	EE/RSH 3366/8409/1963
D 6890–37190–37314	Dalzell	Midland Railway-Butterley	EE/RSH 3368/8411/1964
D 6898–37198§		Dartmoor Railway, Okehampton	EE/RSH 3376/8419/1964
D 6906–37206–37906y§		Severn Valley Railway	EE/VF 3384/D850/1963
D 6907–37207§	William Cooksworthy	Plym Valley Railway	EE/VF 3385/D851/1963
D 6915–37215§		Gloucestershire-Warwickshire Rly	EE/VF 3393/D859/1964
D 6916–37216§	Great Eastern	Pontypool & Blaenavon Railway	EE/VF 3394/D860 1964
D 6919–37219§		Chasewater Railway	EE/VF 3405/D863/1964
D 6927–37227§		Battlefield Railway	EE/VF 3414/D871/1964
D 6940–37240*		Llangollen Railway	EE/VF 3497/D928/1964
D 6954–37254*		Hope Farm, Sellindge	EE/VF 3511/D942/1965
D 6955–37255*		Great Central Railway	EE/VF 3512/D943/1965
D 6963–37263*		Dean Forest Railway	EE/VF 3523/D952/1965
D 6964–37264*		Tyseley Locomotive Works	EE/VF 3524/D953/1965
D 6975–37275*	Stainless Pioneer	Barrow Hill Roundhouse	EE/VF 3535/D964/1965
D 6987–37287–37414	Cathays C&W Works 1846–1993	Weardale Railway	EE/VF 3547/D966/1965

Notes: 37003 also carried the name "First East Anglian Regiment" for a short time in 1963. 37037 also carried the name "GARTCOSH". 37275 also carried the name "Oor Wullie".

37175 is on loan from the Weardale Railway.

CLASS 40 　　　　　　　　　　　　　　　　　　　　　　　　1Co-Co1

Built: 1958–62 by English Electric at Vulcan Foundry, Newton-le-Willows & Robert Stephenson & Hawthorn, Darlington. 200 built.
Engine: English Electric 16SVT MkII of 1480 kW (2000 h.p.) at 850 r.p.m.
Transmission: Electric. Six EE 526/5D axle-hung traction motors.
Power at Rail: 1156 kW (1550 h.p.).
Maximum Tractive Effort: 231 kN (52000 lbf).
Continuous Tractive Effort: 137 kN (30900 lbf) at 18.8 m.p.h.
Weight: 132 tonnes. 　　　　　　　　　**Wheel Diameters:** 914/1143 mm.
Maximum Speed: 90 m.p.h. 　　　　　　**Train Heating:** Steam.
Dual-braked except D 306.

D 200–40122		North Yorkshire Moors Railway (N)	EE/VF 2367/D395 1958
D 212–40012–97407	AUREOL	Midland Railway-Butterley	EE/VF 2667/D429 1959
D 213–40013	ANDANIA	Barrow Hill Roundhouse	EE/VF 2668/D430 1959
D 306–40106	"ATLANTIC CONVEYOR"	Nene Valley Railway	EE/RSH 2726/8136 1960
D 318–40118–97408		Tyseley Locomotive Works	EE/RSH 2853/8148 1961
D 335–40135–97406		East Lancashire Railway	EE/VF 3081/D631 1961
D 345–40145	"East Lancashire Railway"	East Lancashire Railway	EE/VF 3091/D641 1961

CLASS 42 　　　　　　　WARSHIP 　　　　　　　　　　　　　　B-B

Built: 1958–61 at Swindon. 38 Built.
Engines: Two Maybach MD650 of 821 kW (1100 h.p.) at 1530 r.p.m.
Transmission: Hydraulic. Mekydro K 104U.
Maximum Tractive Effort: 223 kN (52400 lbf).
Continuous Tractive Effort: 209 kN (46900 lbf) at 11.5 m.p.h.
Weight: 80 tonnes. 　　　　　　　　　　**Wheel Diameter:** 1033 mm.
Maximum Speed: 90 m.p.h. 　　　　　　**Train Heating:** Steam.

D 821	GREYHOUND	Severn Valley Railway	Swindon 1960
D 832	ONSLAUGHT	East Lancashire Railway	Swindon 1961

CLASS 44 　　　　　　　PEAK 　　　　　　　　　　　　　　　1Co-Co1

Built: 1959–60 at Derby. 10 Built.
Engine: Sulzer 12LDA28A of 1720 kW (2300 h.p.) at 750 r.p.m.
Transmission: Electric. Six Crompton Parkinson C171B1 axle-hung traction motors.
Power at Rail: 1342 kW (1800 h.p.).
Maximum Tractive Effort: 222 kN (50000 lbf).
Continuous Tractive Effort: 129.kN (29100 lbf) at 23.2 m.p.h.
Weight: 135 tonnes. 　　　　　　　　　　**Wheel Diameters:** 914/1143 mm.
Maximum Speed: 90 m.p.h.
Train Heating: Built with steam but facility removed in 1962.

D 4–44004	GREAT GABLE	Midland Railway-Butterley	Derby 1959
D 8–44008	PENYGHENT	Peak Railway	Derby 1959

CLASS 45 — 1Co-Co1

Built: 1960–63 at Crewe and Derby. 127 built.
Engine: Sulzer 12LDA28B of 1860 kW (2500 h.p.) at 750 r.p.m.
Transmission: Electric. Six Crompton Parkinson C172A1 axle-hung traction motors.
Power at Rail: 1491 kW (2000 h.p.).
Maximum Tractive Effort: 245 kN (55000 lbf).
Continuous Tractive Effort: 133 kN (30000 lbf) at 25 m.p.h.
Weight: 135 tonnes. (138 tonnes D14, D53 and D100)
Wheel Diameters: 914/1143 mm. **Maximum Speed:** 90 m.p.h.
Train Heating: Built with steam but subsequently replaced with electric except D 14, D 53 and D 100.
Dual-braked.

D 14–45015		Battlefield Railway	Derby 1960
D 22–45132		Mid Hants Railway	Derby 1961
D 40–45133		Midland Railway-Butterley	Derby 1961
D 53–45041	ROYAL TANK REGIMENT	Midland Railway-Butterley	Crewe 1962
D 61–45112	THE ROYAL ARMY ORDNANCE CORPS	Cotswold Rail, Gloucester	Crewe 1962
D 67–45118	THE ROYAL ARTILLERYMAN	Northampton & Lamport Railway	Crewe 1962
D 86–45105		Barrow Hill Roundhouse	Crewe 1961
D 99–45135	3rd CARABINIER	East Lancashire Railway	Crewe 1961
D 100–45060	SHERWOOD FORESTER	Barrow Hill Roundhouse	Crewe 1961
D 120–45108		LNWR, Crewe Carriage Shed	Crewe 1961
D 123–45125		Great Central Railway	Crewe 1961
D 135–45149	"LEICESTERSHIRE AND DERBYSHIRE YEOMANRY"	Gloucestershire-Warwickshire Rly	Crewe 1961

CLASS 46 — 1Co-Co1

Built: 1961–63 at Derby. 56 built.
Engine: Sulzer 12LDA28B of 1860 kW (2580 h.p.) at 750 r.p.m.
Transmission: Electric. Six Brush TM73-68 MkIII axle-hung traction motors.
Power at Rail: 1460 kW (1960 h.p.).
Maximum Tractive Effort: 245 kN (55000 lbf).
Continuous Tractive Effort: 141 kN (31600 lbf) at 22.3 m.p.h.
Weight: 141 tonnes. **Wheel Diameters:** 914/1143 mm.
Maximum Speed: 90 m.p.h. Dual-braked.
Train Heating: Steam.

D 147–46010	Llangollen Railway	Derby 1961
D 172–46035-97403 IXION	The Railway Age, Crewe	Derby 1962
D 182–46045-97404	Midland Railway-Butterley	Derby 1962

CLASS 47 — Co-Co

Built: 1963–67 at Crewe and Brush Electrical Engineering Company, Loughborough. 512 built.
Engine: Sulzer 12LDA28C of 1920 kW (2580 h.p.) at 750 r.p.m.
Transmission: Electric. Six Brush TG 160-60 axle-hung traction motors.
Power at Rail: 1550 kW (2080 h.p.).
Maximum Tractive Effort: 245 kN (55000 lbf). (267 kN (60000 lbf)§).
Continuous Tractive Effort: 133 kN (33000 lbf) at 26 m.p.h.
Weight: 119–121 tonnes. **Wheel Diameters:** 1143 mm.
Maximum Speed: 95 m.p.h. (100 m.p.h. D 1932 and D1951) Dual-braked.
Train Heating: Built with Steam except D 1787, D 1886 & D 1895 no heat and D 1500, D 1501 & D 1516 steam/electric. Electric heating fitted & steam heating removed on D 1107, D 1566, D 1606, D 1643, D 1656, D 1661, D 1662, D 1723, D 1725, D 1754, D 1755, D 1762, D 1778, D 1909, D 1921, D 1932, D 1933, D 1943, D 1946, D 1951 and D 1970.

D 1705 was built with a Sulzer 12LVA24 engine but replaced with a 12LDA28C in 1972.

D 1107–47524§	Res Gestae	Churnet Valley Railway	Crewe 1966
D 1500–47401	North Eastern	Midland Railway-Butterley	BE 342/1962
D 1501–47402	Gateshead	East Lancashire Railway	BE 343/1962
D 1516–47417		Midland Railway-Butterley	BE 358/1963
D 1524–47004§	Old Oak Common Traction & Rolling Stock Depot	Embsay & Bolton Abbey Railway	BE 419/1963
D 1566–47449§	"ORION"	Llangollen Railway	Crewe 1964
D 1606–47029–47635	The Lass O' Ballochmyle	Swanage Railway	Crewe 1964
D 1643–47059–47631–47765§	Ressaldar	Nottingham Heritage Centre	Crewe 1965
D 1656–47072–47609–47798	FIRE FLY	National Railway Museum	Crewe 1965
D 1661–47077–47613–47840	NORTH STAR	West Somerset Railway	Crewe 1965
D 1662–47484§	ISAMBARD KINGDOM BRUNEL	Rye Farm, Wishaw	Crewe 1965
D 1693–47105§		Gloucestershire-Warwickshire Rly	BE 455/1963
D 1705–47117§	"SPARROWHAWK"	Great Central Railway	BE 467/1965
D 1723–45540§	The Institution of Civil Engineers	Seward Agricultural Machinery, Sinderby	BE 494/1964
D 1725–47490–47768§	Resonant	Barry Island Railway	BE 496/1964
D 1754–47160–47605–47746	The Bobby	St. Modwen Properties, Long Marston	BE 482/1964
D 1755–47541–47773	The Queen Mother	Tyseley Locomotive Works	BE 483/1964
D 1762–47167–47580–47732	Restormel	Tyseley Locomotive Works	BE 524/1964
D 1778–47183–47579–47793	Christopher Wren	Mangapps Farm, Burnham-on-Crouch	BE 540/1964
D 1787–47306	The Sapper	Bodmin Steam Railway	BE 549/1964
D 1842–47192§		Churnet Valley Railway	Crewe 1965
D 1855–47205§		Northampton & Lamport Railway	Crewe 1965
D 1886–47367§		North Norfolk Railway	BE 648/1965
D 1895–47376§	Freightliner 1995	Gloucestershire-Warwickshire Railway	BE 657/1965
D 1909–47232–47665–47785	Fiona Castle	Stainmore Rly, Kirkby Stephen East	BE 671/1965
D 1921–47244–47640§	University of Strathclyde	Battlefield Railway	BE 683/1966
D 1932–47443–47701	Saint Andrew	St. Modwen Properties, Long Marston	BE 694/1966
D 1933–47255–47596§	Aldeburgh Festival	Mid Norfolk Railway	BE 695/1966
D 1943–47500–47770	GREAT WESTERN	Tyseley Locomotive Works	BE 705/1966
D 1946–47503–47771§	Heaton Traincare Depot	Colne Valley Railway	BE 708/1966
D 1951–47507–47716§	Duke of Edinburgh's Award	Dartmoor Railway, Okehampton	BE 613/1966
D 1970–47269–47643§		Bo'ness & Kinneil Railway	Crewe 1965
D 1971–47270	Cory Brothers 1842–1992	Nene Valley Railway	Crewe 1965
D 1994–47292		Nottingham Heritage Centre	Crewe 1966
D 1997–47295§		MoD Ashchurch, Gloucestershire	Crewe 1966

Notes: 47401 also carried the name "Star of the East" for a time. 47701 also carried the names "Waverley" and "Old Oak Common Traction & Rolling Stock Depot". 47732 also carried the name "County of Essex". 47768 also carried the name "Bristol Bath Road". 47770 also carried the name "Reserved". 47771 also carried the name "The Geordie". 47773 also carried the name "Reservist". 47785 also carried the name "The Statesman". 47793 also carried the names "James Nightall GC" and "Saint Augustine". 47798 also carried the name "Prince William".

47798 was also numbered 47834 for a time and 47785 was numbered 47820 for a time.

47798 is currently receiving attention at West Coast Railway Company, Carnforth.

CLASS 50   Co-Co

Built: 1967–68 by English Electric at Vulcan Foundry, Newton-le-Willows. 50 built.
Engine: English Electric 16CVST of 2010 kW (2700 h.p.) at 850 r.p.m.
Transmission: Electric. Six EE 538/5A axle-hung traction motors.
Power at Rail: 1540 kW (2070 h.p.).
Maximum Tractive Effort: 216 kN (48500 lbf).
Continuous Tractive Effort: 147 kN (33000 lbf) at 18.8 m.p.h.
Weight: 117 tonnes.   **Wheel Diameter:** 1092 mm.
Maximum Speed: 100 m.p.h.   Dual-braked.
Train Heating: Electric.

D 400–50050	Fearless	Yeovil Railway Centre	EE/VF 3770/D1141 1967
D 402–50002	Superb	South Devon Railway	EE/VF 3772/D1143 1967
D 407–50007	SIR EDWARD ELGAR	Midland Railway-Butterley	EE/VF 3777/D1148 1968
D 408–50008	Thunderer	LNWR, Crewe Carriage Shed	EE/VF 3778/D1149 1968
D 415–50015	Valiant	East Lancashire Railway	EE/VF 3785/D1156 1968
D 417–50017	Royal Oak	Tyseley Locomotive Works	EE/VF 3787/D1158 1968
D 419–50019	Ramillies	Mid-Norfolk Railway	EE/VF 3789/D1160 1968
D 421–50021	Rodney	Tyseley Locomotive Works	EE/VF 3791/D1162 1968
D 426–50026	Indomitable	RVEL, Derby	EE/VF 3796/D1167 1968
D 427–50027	Lion	North Yorkshire Moors Railway	EE/VF 3797/D1168 1968
D 429–50029	Renown	Peak Railway	EE/VF 3799/D1170 1968
D 430–50030	Repulse	Peak Railway	EE/VF 3800/D1171 1968
D 431–50031	Hood	Severn Valley Railway	EE/VF 3801/D1172 1968
D 433–50033	Glorious	Tyseley Locomotive Works	EE/VF 3803/D1174 1968
D 435–50035	Ark Royal	Severn Valley Railway	EE/VF 3805/D1176 1968
D 440–50040	Centurion	Coventry Railway Centre	EE/VF 3810/D1181 1968
D 442–50042	Triumph	Bodmin Steam Railway	EE/VF 3812/D1183 1968
D 444–50044	Exeter	Severn Valley Railway	EE/VF 3814/D1185 1968
D 449–50049–50149	Defiance	Tyseley Locomotive Works	EE/VF 3819/D1190 1968

Notes: 50007 was previously named "Hercules" and 50040 was previously named "Leviathan".

CLASS 52   WESTERN   C-C

Built: 1961–64 at Crewe and Swindon. 74 built.
Engines: Two Maybach MD655 of 1007 kW (1350 h.p.) at 1500 r.p.m.
Transmission: Hydraulic. Voith L630rV.
Maximum Tractive Effort: 297.3 kN (66770 lbf).
Continuous Tractive Effort: 201.2 kN (45200 lbf) at 14.5 m.p.h.
Weight: 111 tonnes.   **Wheel Diameter:** 1092 mm.
Maximum Speed: 90 m.p.h.   **Train Heating:** Steam.
Dual-braked.

D 1010	WESTERN CAMPAIGNER	West Somerset Railway	Swindon 1962
D 1013	WESTERN RANGER	Severn Valley Railway	Swindon 1962
D 1015	WESTERN CHAMPION	Severn Valley Railway	Swindon 1963
D 1023	WESTERN FUSILIER	National Railway Museum	Swindon 1963
D 1041	WESTERN PRINCE	East Lancashire Railway	Crewe 1962
D 1048	WESTERN LADY	Midland Railway-Butterley	Crewe 1962
D 1062	WESTERN COURIER	Severn Valley Railway	Crewe 1963

Note: D1015 is normally based at Old Oak Common Depot, London.

CLASS 55   DELTIC   Co-Co

Built: 1961–62 by English Electric at Vulcan Foundry, Newton-le-Willows. 22 built.
Engine: Two Napier Deltic T18-25 of 1230 kW (1650 h.p.) at 1500 r.p.m.
Transmission: Electric. Six EE 538 axle-hung traction motors.
Power at Rail: 1969 kW (2640 h.p.).
Maximum Tractive Effort: 222 kN (50000 lbf).
Continuous Tractive Effort: 136 kN (30500 lbf) at 32.5 m.p.h.
Weight: 105 tonnes.   **Wheel Diameter:** 1092 mm.
Maximum Speed: 100 m.p.h.   Dual braked.
Train Heating: Built with Steam, electric subsequently fitted.

D 9000–55022	ROYAL SCOTS GREY	East Lancashire Railway	EE/VF 2905/D557 1961
D 9002–55002	THE KINGS OWN YORKSHIRE LIGHT INFANTRY	National Railway Museum	EE/VF 2907/D559 1961
D 9009–55009	ALYCIDON	North Yorkshire Moors Rly	EE/VF 2914/D566 1961
D 9015–55015	TULYAR	Barrow Hill Roundhouse	EE/VF 2920/D572 1961
D 9016–55016	GORDON HIGHLANDER	Peak Railway	EE/VF 2921/D573 1961
D 9019–55019	ROYAL HIGHLAND FUSILIER	North Yorkshire Moors Rly	EE/VF 2924/D576 1961

CLASS 56 — Co-Co

Built: 1976–84. by Electroputere, Craiova, Romania and BREL Doncaster & Crewe. 135 built.
Engine: Ruston-Paxman 16RK3CT of 2460 kW (3250 h.p.) at 900 r.p.m.
Transmission: Electric. Six Brush TM73-62 axle-hung traction motors.
Power at Rail: 1790 kW (2400 h.p.).
Maximum Tractive Effort: 275 kN (61800 lbf).
Continuous Tractive Effort: 240 kN (53950 lbf) at 32.5 m.p.h.
Weight: 125 tonnes. **Wheel Diameter:** 1143 mm.
Maximum Speed: 80 m.p.h. **Train Heating:** None.
Air braked.

56003	Nene Valley Railway	Electro 752/1976
56022	Weardale Railway	Electro 772/1976
56040 Oystermouth	Mid Norfolk Railway	Doncaster 1978
56057 "British Fuels"	RVEL, Derby	Doncaster 1979
56097	Nottingham Heritage Centre	Doncaster 1981
56098	Northampton & Lamport Railway	Doncaster 1981

CLASS 98/1 — 0-6-0

Built: 1987 by Brecon Mountain Railway. 1 built for use on Aberystwyth–Devil's Bridge line.
Engine: Caterpillar 3304T of 105 kW (140 h.p.).
Transmission: Hydraulic. Twin Disc torque converter.
Gauge: 1' 11½" **Weight:** 12.75 tonnes.
Maximum Speed: 15 m.p.h. **Wheel Diameter:** 610 mm.

10	Vale of Rheidol Railway	BMR 1987

UNCLASSIFIED HUDSWELL-CLARKE — 0-6-0

Built: 1955–61 by Hudswell-Clarke & Company, Leeds. 20 built.
Engine: Gardner 8L3 of 152 kW (204 h.p.) at 1200 r.p.m.
Transmission: Mechanical. SSS powerflow double synchro.
Maximum Tractive Effort: 85.7 kN (19245 lbf).
Continuous Tractive Effort: 76 kN (17069 lbf) at 3.72 m.p.h.
Weight: 34 tonnes. **Wheel Diameter:** 1067 mm.
Maximum Speed: 25 m.p.h.

D 2511	Keighley & Worth Valley Railway	HC D 1202/1961

UNCLASSIFIED NORTH BRITISH — 0-4-0

Built: 1957–61 by North British Locomotive Company, Glasgow. 73 built.
Engine: M.A.N. W6V 17.5/22A of 168 kW (225 h.p.) at 1100 r.p.m.
Transmission: Mechanical. Voith L33YU.
Maximum Tractive Effort: 89.4 kN (20080 lbf).
Continuous Tractive Effort: 53.4 kN (12000 lbf) at 4 m.p.h.
Weight: 28 tonnes. **Wheel Diameter:** 1067 mm.
Maximum Speed: 15 m.p.h.

D 2767	Bo'ness & Kinneil Railway	NBL 28020/1960
D 2774	Strathspey Railway	NBL 28027/1960

2.4. EXPERIMENTAL DIESEL LOCOMOTIVES
PROTOTYPE DELTIC Co-Co
Built: 1955 by English Electric. Used by BR 1959–1961.
Engine: Two Napier Deltic T18-25 of 1230 kW (1650 h.p.) at 1500 r.p.m.
Transmission: Electric. Six EE 526A axle-hung traction motors.
Power at Rail: 1976 kW (2650 h.p.).
Maximum Tractive Effort: 267 kN (60000 lbf).
Continuous Tractive Effort: 104 kN (23400 lbf) at 43.5 m.p.h.
Weight: 107.7 tonnes. **Wheel Diameter:** 1092 mm.
Maximum Speed: 105 m.p.h. **Train Heating:** Steam.

DELTIC   Locomotion: The NRM at Shildon   EE 2003/1955

PROTOTYPE ENGLISH ELECTRIC SHUNTER 0-6-0
Built: 1957 by English Electric at Vulcan Foundry, Newton-le-Willows. Used by BR 1957–1960.
Engine: English Electric 6RKT of 373 kW (500 h.p.) at 750 r.p.m.
Transmission: Electric.
Power at Rail:
Maximum Tractive Effort: 147 kN (33000 lbf).
Continuous Tractive Effort: (lbf) at m.p.h.
Weight: 48 tonnes. **Wheel Diameter:** 1219 mm.
Maximum Speed: 35 m.p.h.

D 226–D 0226   "VULCAN"   Keighley & Worth Valley Railway   EE/VF 2345/D226 1956

PROTOTYPE NORTH BRITISH SHUNTERS 0-4-0
Built: 1954 by North British Locomotive Company, Glasgow. Used by BR Western Region (27414) and BR London Midland & Southern regions (27415). Subsequently sold for industrial use.
Engine: Paxman 6 VRPHXL of 160 kW (225 h.p.) at 1250 r.p.m.
Transmission: Hydraulic. Voith L24V. **Weight:** tonnes.
Maximum Tractive Effort: 112 kN (22850 lbf). **Max. Speed:** 12 m.p.h.
Wheel Diameter: 1016 mm. No train brakes.

BR	Present		
–	TOM	Telford Steam Railway	NBL 27414/1954
–	TIGER	Bo'ness & Kinneil Railway	NBL 27415/1954

PROTOTYPE NORTH BRITISH SHUNTER 0-4-0
Built: 1958 by North British Locomotive Company, Glasgow. Used by BR Western Region at Old Oak Common in 1958. Subsequently sold for industrial use.
Engine: MAN W6V 17.5/22 OF 168 kW (225 h.p.).
Transmission: Hydraulic. Voith L24V. **Weight:** tonnes.
Maximum Tractive Effort: 112 kN (22850 lbf). **Maximum Speed:** 12 m.p.h.
Wheel Diameter: 940 mm. No train brakes.

BR	Present		
–	D1	The Pallot Heritage Steam Museum, Jersey	NBL 27734/1958

2.5. CIVIL ENGINEERS DIESEL LOCOMOTIVES
CLASS 97/6 0-6-0
Built: 1952–59 by Ruston & Hornsby at Lincoln for BR Western Region Civil Engineers. 5 built.
Engine: Ruston 6VPH of 123 kW (165 h.p.).
Transmission: Electric. One British Thomson Houston RTA5041 traction motor.
Maximum Tractive Effort: 75 kN (17000 lbf). **Weight:** 31 t.
Maximum Speed: 20 m.p.h. **Wheel Diameter:** 978 mm.

PWM 650–97650	Lincolnshire Wolds Railway	RH 312990/1952
PWM 651–97651	Northampton & Lamport Railway	RH 431758/1959
PWM 653–97653	Long Marston Workshops	RH 431760/1959
PWM 654–97654	Peak Railway	RH 431761/1959

PLATFORM 5 MAIL ORDER

PLATFORM 5 EUROPEAN HANDBOOKS

The Platform 5 European Handbooks are the most comprehensive guides to the rolling stock of selected European railway administrations available. Each book in the series contains the following information:

- Locomotives
- Railcars and Multiple Units
- Depot Allocations (where allocated)
- Preserved Locomotives, Museums and Museum Lines
- Technical Data
- Lists of Depots and Workshops

Each book is A5 size, thread sewn and includes at least 32 pages of colour photographs. Benelux Railways also includes details of hauled coaching stock. The following are currently available:

No.2A: German Railways 4th edition Part 1: DB Locos & MUs .... £16.95   176 pages. Published 2004
No.2B: German Railways 4th edition Part 2:
   Private Railways & Museums .... £16.95   224 pages. Published 2004
No.3: Austrian Railways 4th edition .... £17.50   176 pages. Published 2005
No.4: French Railways 3rd edition .... £14.50   176 pages. Published 1999
No.6: Italian Railways 2nd edition .... £18.50   192 pages. Published 2007
No.7: Irish Railways 2nd edition .... £12.95   96 pages.   Due late 2007

If you would like to be notified when new titles in this series are published, please contact our Mail Order Department. Alternatively, please see our advertisements in **Today's Railways UK** or **Today's Railways Europe** magazines for up to date publication information.

HOW TO ORDER

Telephone your order and credit/debit card details to our 24-hour sales orderline:
0114 255 8000 or Fax: 0114 255 2471.
An answerphone is attached for calls made outside of normal UK office hours.
Or send your credit/debit card details, sterling cheque, money order or British Postal order payable to 'Platform 5 Publishing Ltd.' to:

Mail Order Department (PL), Platform 5 Publishing Ltd, 3 Wyvern House, Sark Road, SHEFFIELD, S2 4HG, ENGLAND

Please add postage & packing: 10% UK; 20% Europe; 30% Rest of World.
Please allow 28 days for delivery in the UK.

3. ELECTRIC LOCOMOTIVES

Electric railways have existed in Great Britain for over one hundred years. Prior to the Second World War the majority of electrification was for the movement of passengers in and to metropolitan areas. The North Eastern Railway did however build a small fleet of electric locomotives for hauling heavy coal and steel trains in County Durham.

Notes
For notes on wheel arrangements, dimensions, tractive effort and brakes see "Diesel Locomotives" section.

3.1. PRE-NATIONALISATION DESIGN ELECTRIC LOCOMOTIVES

LSWR   Bo
Built: 1898. Siemens design for operation on the Waterloo & City line.
System: 750 V DC third rail.   **Train Heating:** None
Traction Motors: Two Siemens 45 kW (60 h.p).
Wheel dia.: 3' 4".

BR	SR	LSWR		
DS75	75S	Unnumbered	Midleton Railway (N)	SM 6/1898

NORTH EASTERN RAILWAY CLASS ES1   Bo-Bo
Built: 1905. 2 built. Used on Newcastle Riverside Branch.
System: 600 V DC overhead.   **Train Heating:** None
Traction Motors: 4 BTH design.
Weight: 46 tonnes.   **Wheel Dia:** 915 mm.

BR	LNER	NER		
26500	4075–6480	1	Locomotion: The NRM at Shildon	BE1905

LNER CLASS EM1 (BR CLASS 76)   Bo+Bo
Built: 1941–53 at Doncaster (26000) and Gorton (others) for Manchester–Sheffield/Wath system. 58 built.
System: 1500 V DC overhead.
Traction Motors: 4 MV 186 axle-hung.
Max Rail Power: 2460 kW (3300 h.p).
Continuous Rating: 970 kW (1300 h.p)
Max. T.E.: 200 kN (45000 lbf).   **Weight:** 88 tonnes.
Cont. T.E.: 39 kN (8800 lbf) at 56 m.p.h.   **Wheel Dia:** 1270 mm.
Max. Speed: 65 m.p.h.   **Train Heating:** Steam

26020–E26020–76020	National Railway Museum	Gorton 1027/1951

LNER CLASS EM2 (BR CLASS 77)   Co-Co
Built: 1953–55 at Gorton for BR to LNER design for Manchester–Sheffield route. 7 built. Sold to NS (Netherlands Railways) 1969.
System: 1500 V DC overhead.
Traction Motors: 6 MV 146 axle-hung.
Max Rail Power: 1716 kW (2300 hp).
Max. T.E.: 200 kN (45000 lbf).   **Weight:** 102 tonnes.
Cont. T.E.: 78 kN (15600 lbf) at 23 m.p.h. **Wheel Dia:** 1092 mm.
Max. Speed: 90 m.p.h.
Train Heating: Steam whilst on BR, electric fitted by NS.
Air brakes.

NS	BR				
1502	27000–E 27000	ELECTRA	Midland Railway-Butterley	Gorton 1065/1953
1505	27001–E 27001	ARIADNE	Greater Manchester Museum of Science & Industry	Gorton 1066/1954
1501	27003–E 27003	(DIANA)	Leidseddamm, Den Haag (NS)	Gorton 1068/1954

3.2. BRITISH RAILWAYS ELECTRIC LOCOMOTIVES

Numbering System

Numbering of electric locomotives from 1957 was similar to that of diesel locomotives, except that the numbers were prefixed with an "E" instead of a "D". Locomotives of pre-nationalisation design continued to be numbered in the 2xxxx series, although Classes EM1 and EM2 later acquired an "E" prefix to their existing numbers. As with diesels, electric locomotives were later allocated a two-digit class number followed by a three-digit serial number.

CLASS 71 Bo-Bo

Built: 1958–60 at Doncaster. 24 built.
System: 660–750 V DC third rail or overhead.
Continuous Rating: 1715 kW (2300 h.p.).
Max. T.E.: 191 kN (43 000 lbf).
Cont. T.E.: 55 kN (12 400 lbf) at 69.6 m.p.h.
Max. Speed: 90 m.p.h.
Weight: 76.2 tonnes.
Wheel Dia: 1219 mm.
Train Heating: Electric
Dual braked.

E 5001–71001   Locomotion: The NRM at Shildon   Doncaster 1959

CLASS 73/0 ELECTRO-DIESEL Bo-Bo

Built: 1962 at Eastleigh. 6 built.
System: 660–750 V DC from third rail.   **Continuous Rating:** Electric 1060 kW (1420 h.p.).
Continuous Tractive Effort: Diesel 72 kN (16100 lbf) at 10 m.p.h.
Maximum Tractive Effort: Electric 187 kN (42000 lbf). Diesel 152 kN (34100 lbf).
Weight: 76.3 tonnes.   **Wheel Dia:** 1016 mm.
Max. Speed: 80 m.p.h.
Triple braked (vacuum, air and electro-pneumatic).
Train Heating: Electric

E 6001–73001–73901		Dean Forest Railway	Eastleigh 1962
E 6002–73002		Dean Forest Railway	Eastleigh 1962
E 6003–73003	Sir Herbert Walker	Swindon & Cricklade Railway	Eastleigh 1962
E 6005–73005		Severn Valley Railway	Eastleigh 1962
E 6006–73006–73906		Severn Valley Railway	Eastleigh 1962

CLASS 73/1 ELECTRO-DIESEL Bo-Bo

Built: 1965–67 by English Electric at Vulcan Foundry, Newton-le-Willows. 43 built.
System: 660–750 V DC from third rail.   **Continuous Rating:** Electric 1060 kW (1420 h.p.).
Continuous Tractive Effort: Diesel 60 kN (13600 lbf) at 11.5 m.p.h.
Maximum Tractive Effort: Electric 179 kN (40000 lbf). Diesel 152 kN (34100 lbf).
Weight: 76.8 tonnes.   **Wheel Dia:** 1016 mm.
Max. Speed: 90 m.p.h.
Triple braked (vacuum, air and electro-pneumatic).
Train Heating: Electric.

E 6007–73101	The Royal Alex'	Severn Valley Railway	EE/VF 3569/E339 1965
E 6010–73104		Allely's, The Slough, Studley	EE/VF 3572/E342 1965
E 6011–73105	Quadrant	Battlefield Railway	EE/VF 3573/E343 1966
E 6016–73110		Churnet Valley Railway	EE/VF 3578/E348 1966
E 6020–73114	Stewarts Lane Traction & Maintenance Depot	Battlefield Railway	EE/VF 3582/E352 1966
E 6025–73119	Kentish Mercury	Keith & Dufftown Railway	EE/VF 3587/E357 1966
E 6033–73126	"Kent & East Sussex Railway"	Fire Service College, Moreton-in-Marsh	EE/VF 3595/E365 1966
E 6035–73128	OVS BULLEID C.B.E.	Chasewater Railway	EE/VF 3597/E367 1966

E 6036–73129	City of Winchester	Gloucestershire-Warwickshire Rly	EE/VF 3598/E368 1966
E 6041–73134	Woking Homes 1885-1995	Weardale Railway	EE/VF 3713/E373 1966
E 6045–73138	Poste Haste 150 YEARS OF TRAVELLING POST OFFICES		
		St. Modwen Properties, Long Marston	EE/VF 3717/E377 1966
E 6046–73139		Barrow Hill Roundhouse	EE/VF 3718/E378 1966
E 6047–73140		Spa Valley Railway	EE/VF 3719/E379 1966

Notes: 73101 also carried the name "Brighton Evening Argos".

73101 and 73105 were numbered 73801 and 73805 for a time.

CLASS 81   Bo-Bo

Built: 1959–64 by Birmingham R.C.&W Company, Birmingham. 25 Built.
System: 25 kV AC overhead.   **Continuous Rating:** 2390 kW (3200 h.p.).
Max. T.E.: 222 kN (50000 lbf).   **Weight:** 79 tonnes.
Cont. T.E.: 76 kN (17000 lbf) at 71 m.p.h.   **Wheel Dia:** 1219 mm.
Max. Speed: 100 m.p.h.   **Train Heating:** Electric
Dual braked.

E3003–81002   Barrow Hill Roundhouse   BTH 1085/1960

CLASS 82   Bo-Bo

Built: 1960–62 by Beyer Peacock, Manchester. 10 Built.
System: 25 kV AC overhead.   **Continuous Rating:** 2460 kW (3300 h.p.).
Max. T.E.: 222 kN (50000 lbf).   **Weight:** 80 tonnes.
Cont. T.E.: 76 kN (17000 lbf) at 73 m.p.h.   **Wheel Dia:** 1219 mm.
Max. Speed: 100 m.p.h.   **Train Heating:** Electric
Dual braked.

E3054–82008   Barrow Hill Roundhouse   BP 7893/1961

CLASS 83   Bo-Bo

Built: 1960–62 by English Electric at Vulcan Foundry, Newton-le-Willows. 15 Built.
System: 25 kV AC overhead.   **Continuous Rating:** 2200 kW (2950 h.p.).
Max. T.E.: 169 kN (38000 lbf).   **Weight:** 76 tonnes.
Cont. T.E.: 68 kN (15260 lbf) at 73 m.p.h.   **Wheel Dia:** 1219 mm.
Max. Speed: 100 m.p.h.   **Train Heating:** Electric
Dual braked.

E3035–83012   Barrow Hill Roundhouse   EE 2941/VF E277/1961

CLASS 84   Bo-Bo

Built: 1960–61 by North British Locomotive Company, Glasgow. 10 built.
System: 25 kV AC overhead.   **Continuous Rating:** 2312 kW (3100 h.p.).
Max. T.E.: 222 kN (50000 lbf).   **Weight:** 76.6 tonnes.
Cont. T.E.: 78 kN (17600 lbf) at 66 m.p.h.   **Wheel Dia:** 1219 mm.
Max. Speed: 100 m.p.h.   **Train Heating:** Electric
Dual braked.

E 3036–84001   Barrow Hill Roundhouse (N)   NBL 27793/1960

CLASS 85   Bo-Bo

Built: 1961–65 at Doncaster. 40 Built.
System: 25 kV AC overhead.   **Continuous Rating:** 2390 kW (3200 h.p.).
Max. T.E.: 222 kN (50000 lbf).   **Weight:** 82.5 tonnes.
Cont. T.E.: 76 kN (17000 lbf) at 71 m.p.h.   **Wheel Dia:** 1219 mm.
Max. Speed: 100 m.p.h.   **Train Heating:** Electric
Dual braked.

E3061–85006–85101   "Doncaster Plant 150 1853–2003"
                           Barrow Hill Roundhouse   Doncaster 1961

CLASS 86 — Bo-Bo

Built: 1965–66 at Doncaster or English Electric Company at Vulcan Foundry, Newton-le-Willows. 100 built.
System: 25 kV AC overhead.
Continuous Rating: 2680 kW (3600 h.p.) (* 5860 kW (5000 h.p.) († 3010 kW (4040 h.p.).
Continuous Tractive Effort: 89 kN (20 000 lbf) (* 95 kN 21 300 lbf) († 85 kN 19 200 lbf).
Maximum Tractive Effort: 258 kN (58 000 lbf) († 207 kN (46 500 lbf)
Weight: 83 tonnes (* 87 tonnes) († 85 tonnes).
Wheel Dia: 1156 mm (* 1150 mm).
Max. Speed: 100 m.p.h. (* 110 m.p.h.)   **Train Heating:** Electric.

E 3137–86045–86259†	Les Ross	Tyseley Locomotive Works	Doncaster 1966
E 3191–86201–86101*	Sir William A Stanier FRS	LNWR, Crewe Carriage Shed	EE/VF3483/E329 1965
E 3193–86213†	Lancashire Witch	Wembley Depot, London	EE/VF3485/E331 1965
E 3194–86001–86401	Northampton Town	St. Modwen Properties, Long Marston	EE/VF3491/E337 1966

Note: E 3137 was originally named "Peter Pan".

CLASS 87 — Bo-Bo

Built: 1975–76 by BREL at Crewe Works. 36 built.
System: 25 kV AC overhead.
Continuous Rating: 3730 kW (5000 h.p.).
Continuous Tractive Effort: 95 kN (21 300 lbf) at 87 m.p.h.
Maximum Tractive Effort: 258 kN (58 000 lbf).
Weight: 83.5 tonnes.
Wheel Dia: 1150 mm.
Max. Speed: 110 m.p.h.   **Train Heating:** Electric. Air braked

87001	Royal Scot	National Railway Museum	Crewe 1973
87012	Coeur-de-Lion	BZK, Bulgaria	Crewe 1973
87019	Sir Winston Churchill	BZK, Bulgaria	Crewe 1973
87031	Hal o' the Wynd	Tyseley Locomotive Works	Crewe 1974
87035	Robert Burns	The Railway Age, Crewe	Crewe 1974

Note: 87001 also carried the name "STEPHENSON". 87012 also carried the name "The Royal Bank of Scotland for a time" and was later named "The Olympian". 87019 was later named "ACoRP Association of Community Rail Partnerships".

CLASS 89 — Co-Co

Built: 1987 by BREL at Crewe Works. 1 built.
System: 25 kV AC overhead.
Continuous Rating: 4350 kW (6550 h.p.).
Continuous Tractive Effort: 105 kN (23 600 lbf) at 92 m.p.h.
Maximum Tractive Effort: 205 kN (46 000 lbf).
Weight: 104 tonnes.
Wheel Dia: 1150 mm.
Max. Speed: 125 m.p.h.   **Train Heating:** Electric. Air braked

89001	Avocet	Barrow Hill Roundhouse	Crewe 1987

4. GAS TURBINE VEHICLES

LOCOMOTIVE AIA-AIA
Built: 1950 by Brown Boveri in Switzerland.
Power Unit: Brown Boveri gas turbine of 1828 kW (2450 h.p.).
Transmission: Electric. Four traction motors.
Max. T.E.: 140 kN (31500 lbf). **Weight:** 117.1 tonnes.
Cont. T.E.: 55 kN (12400 lbf) at 64 m.p.h. **Wheel Dias:** 1232 mm.
Max. Speed: 90 m.p.h. **Train Heating:** Steam.

18000	The Railway Age, Crewe	BBC 4559/1950

EXPERIMENTAL ADVANCED PASSENGER TRAIN (APT-E)
Built: 1972 at Derby Litchurch Lane Works.
Power Units: Eight Leyland 350 automotive gas turbines of 222 kW (298 h.p.).
Traction Motors: Four GEC 253AY. Articulated unit.

PC1	Locomotion: The NRM at Shildon	Derby 1972
PC2	Locomotion: The NRM at Shildon	Derby 1972
TC1	Locomotion: The NRM at Shildon	Derby 1972
TC2	Locomotion: The NRM at Shildon	Derby 1972

PLATFORM 5 MAIL ORDER

RAIL ATLAS OF GREAT BRITAIN & IRELAND 11th edition
Oxford Publishing Company
Fully revised and updated edition of the definitive UK railway atlas. Shows all lines with multiple/single track distinction and colours used to denote municipal/urban railways, preserved lines and freight only lines. Also shows all stations, junctions, freight terminals, depots, works and marshalling yards. Includes many enlargements of complicated layouts, plus an index map and a full index to locations. 128 pages. Hardback. £14.99

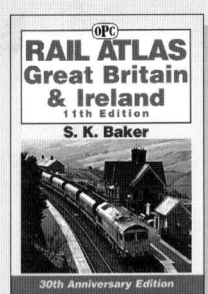

BRITISH MULTIPLE UNITS VOL 1: DMUs & DEMUs 2nd edition
Coorlea Publishing
Fully revised and updated reference work listing all DMU and DEMU vehicles that have operated on Britain's railways. Contains details of all first and second generation vehicles including new information and alterations that have occurred since the first edition was published in 2000, plus new unit additions such as Class 185s. Contains date of entry to service, set formations, formation changes, renumbering, withdrawal and disposal information for all vehicles. Very detailed. 84 pages. £14.95.

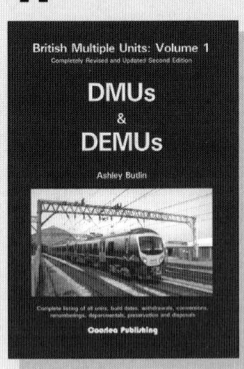

ALSO AVAILABLE IN THE SAME SERIES:
Volume 2: EPBs, Haps, Caps & Saps £6.95
Volume 3: Classes 302-390 .... £9.95
Volume 4: Classes 410-490 & 508 .... £13.95

HOW TO ORDER
Telephone your order and credit/debit card details to our 24-hour sales orderline:
0114 255 8000 or Fax: 0114 255 2471.
An answerphone is attached for calls made outside of normal UK office hours.
Or send your credit/debit card details, sterling cheque, money order or British Postal order payable to 'Platform 5 Publishing Ltd.' to:

Mail Order Department (PL), Platform 5 Publishing Ltd, 3 Wyvern House, Sark Road, SHEFFIELD, S2 4HG, ENGLAND
Please add postage & packing: 10% UK; 20% Europe; 30% Rest of World.
Please allow 28 days for delivery in the UK.

▲ An impressive line up of six shunting locomotives at Rowsley, Peak Rail on 17 September 2005. Leading is 03099, followed by 07013, 08016, Class 02 D2854, Class 05 D2587 and Class 14 D9525.
Chris Booth

▼ Class 12 15224 is seen at Tunbridge Wells West on the Spa Valley Railway on 10 August 2003.
Alan Barnes

▲ There are four Class 14s based at the Nene Valley Railway. Here D9520 leads D9523 and D9516 at Ferry Meadows with the 10.50 Wansford–Peterborough NV on 3 March 2007.

▼ Class 25 D7612 (25262) is seen between Groombridge and High Rocks on the Spa Valley Railway with the 17.32 Groombridge Junction–Tunbridge Wells West brake van ride on 4 August 2007.

Robert Pritchard (2)

115

In Railfreight grey livery 20118 "Saltburn-by-the-Sea" passes Riverford Bridge on the South Devon Railway with the 13.00 Buckfastleigh–Totnes on 9 June 2007. **Phil Chilton**

▲ Class 27 D5401 is seen at Thurcaston, having just left Leicester North on the Great Central Railway with the 17.15 Leicester North–Loughborough Central on 28 April 2007. **Robert Pritchard**

▼ Also on the Great Central golden ochre-liveried Class 31 D5830 is seen at Woodthorpe with the 14.30 Loughborough Central–Rothley on 7 July 2007. **Paul Biggs**

▲ The Swanage Railway's Class 33 D6515 "Stan Symes" leaves Corfe Castle with the 11.50 Norden–Swanage on 27 November 2005. **Andrew Mist**

▼ "Hymek" D7017 is seen stabled at Williton shed on the West Somerset Railway on 1 October 2005. **Chris Wilson**

118

37901 "Mirrlees Pioneer" powers a photographer's freight charter at Garthydawr on the Llangollen Railway on 17 June 2006. **Paul Robertson**

▲ Class 40 D306 "ATLANTIC CONVEYOR" leaves Wansford with the 10.35 to Peterborough NV on 9 October 2004.   **Robert Pritchard**

▼ "Warship" D832 ONSLAUGHT awaits departure from Ramsbottom on the rear of the 15.05 service to Rawtenstall on 6 July 2006.   **Paul Biggs**

Class 47 D1842 (47192) pauses under the former loading hopper at Oakamoor (used for loading sand from the adjacent quarry), on the Churnet Valley Railway on 2 July 2007, during an EMRPS "Staffordshire Brush" photo charter.
Laurence Sly

121

▲ 46010 is seen at Hendon on the delightful Llangollen Railway with the 14.35 Llangollen–Carrog on a seasonal 29 February 2004. **Paul Robertson**

▼ Main line regular 50049 Defiance works the 16.45 Minehead–London Victoria "Cathedrals Express" charter on the West Somerset Railway near Williton on 18 July 2007. **Paul Waring**

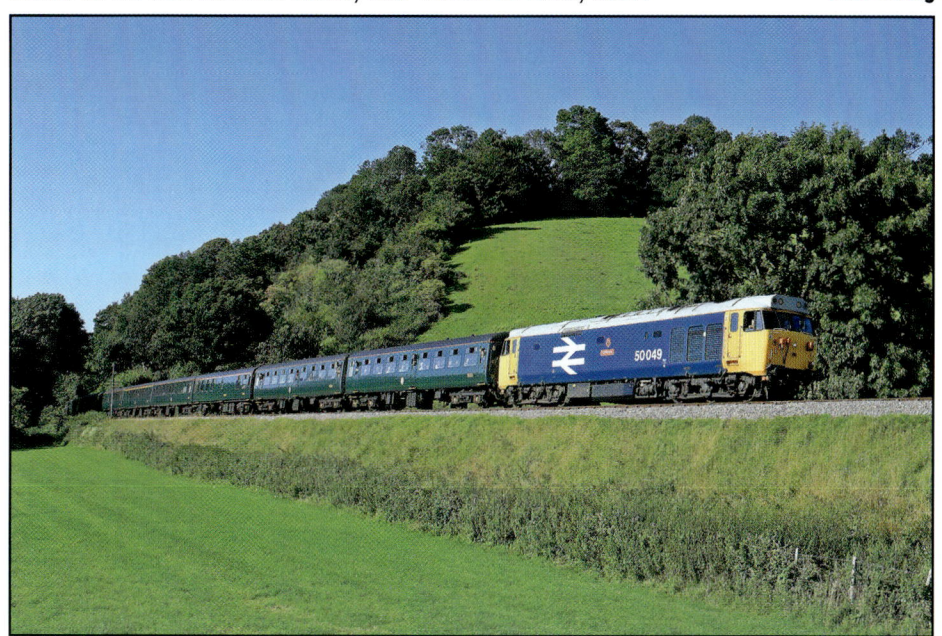

▲ Recently repainted into BR Maroon livery, D1015 WESTERN CHAMPION passes Burrs on the East Lancashire Railway with the 14.26 Rawtenstall–Heywood on 7 July 2007. The loco was visiting the line and is normally based at Old Oak Common depot, London. **Phil Chilton**

▼ DELTIC D9009 ALYCIDON leaves Levisham on the North Yorkshire Moors Railway with the 18.00 Pickering–Grosmont on 28 May 2007. **Andrew Mason**

123

In glorious evening light the Spa Valley Railway's Class 73 D6047 (73140) passes Pokehill Farm Crossing, between High Rocks and Groombridge, with the 19.30 Tunbridge Wells West–Birchden Junction on 4 August 2007. **Robert Pritchard**

▲ Several Class 56s are now preserved. 56003, in Loadhaul livery, and visiting 56098 in Trainload Grey cross the River Nene at Wansford with the 13.48 from Peterborough NV on 3 March 2007.
Andrew Mist

▼ 86101 Sir William A Stanier FRS passes Minshull Vernon near Crewe with a Compass Railtour from Carlisle to Crewe (then Holyhead) on 24 March 2007. **Hugh Ballantyne**

▲ Newly repainted in Inter-City livery, 89001 is seen at Barrow Hill Roundhouse on 7 July 2007. This loco is one of several electric locos owned by the AC Locomotive Group. **Mick Tindall**

▼ Class 101 DMU (vehicles DMCL 50321+DMBS 51427) is seen at the superbly restored Quorn & Woodhouse station on the Great Central Railway on 4 November 2006 with the 09.45 Loughborough Central–Leicester North service. **Paul Chancellor**

▲ The Llangollen Railway's Blue & Grey Class 108 (DMBS 51907+DTCL 54490) is seen at Llangollen on 10 June 2006. **Paul Chancellor**

▼ Restored Unclassified Derby Lightweight single car 79900 approaches Swanwick Junction on the Midland Railway-Butterley with the 13.30 from Butterley on 16 July 2006. This vehicle was formerly in departmental use. **Chris Booth**

▲ Class 205 "Thumper" DEMU 205 018 (60117+60828) is resident at the Pontypool & Blaenavon Railway in South Wales. On 30 July 2006 it approaches Furnace Halt, Blaenavon with the 14.40 from Whistle Halt. **Andrew Mist**

▼ The Great Central Railway have a 4-car Class 421 "Cig" EMU. Here 1393 (76817, 70527, 62384 and 76746) is seen stabled at Swithland on 4 November 2006. **Paul Chancellor**

▲ The prototype "DELTIC" is seen at Barrow Hill on 5 October 2003. **Robert Pritchard**

▼ LNER design Class EM2 27000 "ELECTRA" is seen on display at Barrow Hill Roundhouse on 6 October 2004. **Alan Barnes**

5. DIESEL MULTIPLE UNIT VEHICLES

GENERAL

Prior to nationalisation, several schemes to transfer road bus technology to rail vehicles had taken place with little success, the exception being the GWR where the concept was developed resulting in a fleet of distinctive diesel railcars.

During the 1950s British Railways took the idea far more seriously as part of the modernisation plan and consequently numerous designs appeared. The preservation of these has expanded enormously in recent years.

Type Codes

The type codes used by the former BR operating departments to describe the various types of multiple unit vehicles are used, these being:

B   Brake, i.e. a vehicle with luggage space and a guard's/conductor's compartment.

BDM	Battery driving motor	M	Motor
BDT	Battery driving trailer	O	Open vehicle
C	Composite	P	Pullman
Cso	Semi-open composite	PMV	Parcels & miscellaneous van
DM	Driving motor	RB	Buffet car
DT	Driving trailer	S	Second (now known as standard)
F	First	Sso	Semi-open second
K	Side corridor with lavatory	T	Third (reclassified second in 1956)
L	Open or semi-open with lavatory	T	Trailer
LV	Luggage Van		

All diesel mechanical and diesel hydraulic vehicles are assumed to be open unless stated otherwise and do not carry an "O" in the code. Certain DMU vehicles are nowadays used as hauled stock and these are denoted by a letter "h" after the number.

Prefixes and Suffixes

Coaching stock vehicles used to carry regional prefix letters to denote the owning region. These were removed in the 1980s. These are not shown. Pre-nationalisation number series vehicles carried both prefix and suffix letters, the suffix denoting the pre-nationalisation number series. The prefixes and suffixes are shown for these vehicles.

Dimensions

Dimensions are shown as length (over buffers or couplers) x width (over bodysides including door handles).

Seating Capacities

These are shown as nF/nS relating to first and second class seats respectively, e.g. a car with 12 first class seats and 51 second class seats would be shown as 12/51. Prior to 3rd June 1956 second class was referred to as "third" class and is now referred to as "standard" class. Certain old vehicles are thus shown as third class.

Bogies

All vehicles are assumed to have two four-wheeled bogies unless otherwise stated.

5.1. GWR DIESEL RAILCARS

UNCLASSIFIED — PARK ROYAL
Built: 1934 by Park Royal. Single cars with two driving cabs.
Engines: Two AEC 90 kW (121 h.p.). **Transmission:** Mechanical.
Body: 19.58 x 2.70 m. **Weight:** 26.6 tonnes. **Seats:** –/44.
Max. Speed: 75 m.p.h.

BR	GWR		
W 4 W	4	Steam – Museum of the Great Western Railway, Swindon (N)	PR 1934

UNCLASSIFIED — GWR
Built: 1940 at Swindon. Single cars with two driving cabs.
Engines: Two AEC 78 kW (105 h.p.). **Transmission:** Mechanical.
Body: 20.21 x 2.70 m. **Weight:** 36.2 tonnes. **Seats:** –/48.
Max. Speed: 40 m.p.h.

BR	GWR		
W 20 W	20	Kent & East Sussex Railway	Swindon 1940
W 22 W	22	Didcot Railway Centre	Swindon 1941

5.2. BRITISH RAILWAYS DMUs

Numbering System
Early BR diesel multiple units were numbered in the 79xxx series, but when it was evident that this series did not contain enough numbers the 5xxxx series was allocated to this type of vehicle and the few locomotive-hauled non-corridor coaches which were in the 5xxxx series were renumbered into the 4xxxx series. Power cars in the 50xxx series and driving trailers in the 56xxx series were eventually renumbered into the 53xxx and 54xxx series respectively to avoid conflicting numbers with Class 50 and 56 diesel locomotives.

Diesel-electric multiple unit power cars were renumbered in the 60000–60499 series, trailers in the 60500–60799 series and driving trailers in the 60800–60999 series.

5.2.1. HIGH SPEED DIESEL TRAINS
The prototype High Speed Diesel Train (HSDT) appeared in 1972. The two power cars (41001/2) were constructed to a locomotive lot (No. 1501) whilst the intermediate vehicles (10000, 10100, 11000–11002 and 12000–12002) were constructed to coaching stock lots. On 10 July 1974 the power cars were reclassified as coaching stock and issued with a coaching stock lot number. All prototype HSDT vehicles were thus categorised as multiple unit stock and renumbered into the 4xxxx series, Class No. 252 being allocated for the complete set.

CLASS 252 — PROTOTYPE HST POWER CAR
Built: 1972 at Derby. (2 built).
Engine: Paxman Valenta 12RP200L of 1680 kW (2250 h.p.) at 1500 r.p.m.
Traction Motors: Four Brush TMH 68-46.
Power at Rail: 1320 kW (1770 h.p.).
Max. T.E.: 80 kN (17 980 lbf). **Weight:** 67 tonnes
Cont. T.E.: 46 kN (10 340 lbf.) at 64.5 m.p.h. **Wheel Dia:** 1020 mm.
Max. Speed: 125 m.p.h. Air braked.

41001–43000–ADB 975812	National Railway Museum	Derby 1972

5.2.2. DIESEL MECHANICAL MULTIPLE UNITS
Note: All vehicles in this section have a maximum speed of 70 m.p.h.

CLASS 100 — GRCW TWIN UNITS
Built: 1957–58. Normal formation: DMBS–DTCL.
Engines: Two AEC 220 of 112 kW (150 h.p.).

DMBS	18.49 x 2.82 m	30.5 tonnes	–/52	
DTCL	18.49 x 2.82 m	25.5 tonnes	12/54	

51118	DMBS	Midland Railway-Butterley	GRCW 1957
56097	DTCL	Midland Railway-Butterley	GRCW 1957
56301	DTCL	Mid Norfolk Railway	GRCW 1957
56317	DTCL	Essex Traction Group, Boxted	GRCW 1958

CLASS 101 — METRO-CAMMELL UNITS
Built: 1958–59. Various Formations.
Engines: Two AEC 220 of 112 kW (150 h.p.).

DMBS	18.49 x 2.82 m	32.5 tonnes	–/52	
DMCL	18.49 x 2.82 m	32.5 tonnes	12/46 (originally 12/53)	
DTCL	18.49 x 2.82 m	25.5 tonnes	12/53	
TCL	18.49 x 2.82 m	25.5 tonnes	12/53	
TSL	18.49 x 2.82 m	25.5 tonnes	–/71	

50160–53160	DMCL	Chasewater Light Railway	MC 1956
50164–53164	DMBS	Chasewater Light Railway	MC 1957
50170–53170	DMCL	Midland Railway-Butterley	MC 1957
50193–53193	DMCL	Great Central Railway	MC 1957
50203–53203	DMBS	Great Central Railway	MC 1957
50204–53204	DMBS	North Yorkshire Moors Railway	MC 1957
50253–53253	DMBS	Midland Railway-Butterley	MC 1957
50256–53256	DMBS	East Kent Light Railway	MC 1957
50266–53266	DMCL	Great Central Railway	MC 1957
50268–53268	DMCL	Keighley & Worth Valley Railway	MC 1957
50321–53321–977900	DMCL	Great Central Railway	MC 1958
50746–53746	DMCL	Wensleydale Railway	MC 1957
51187	DMBS	Cambrian Railway Trust, Llynclys	MC 1958
51188	DMBS	Ecclesbourne Valley Railway	MC 1958
51189	DMBS	Keighley & Worth Valley Railway	MC 1958
51192	DMBS	East Lancashire Railway (N)	MC 1958
51205	DMBS	Cambrian Railway Trust, Llynclys	MC 1958
51210	DMBS	Wensleydale Railway	MC 1958
51213	DMBS	East Anglian Railway Museum	MC 1958
51226	DMBS	Mid Norfolk Railway	MC 1958
51228	DMBS	North Norfolk Railway	MC 1958
51247	DMBS	Wensleydale Railway	MC 1958
51427–977899	DMBS	Great Central Railway	MC 1959
51434 "Matthew Smith"	DMBS	Mid Norfolk Railway	MC 1959
51499	DMCL	Mid Norfolk Railway	MC 1959
51503	DMCL	Mid Norfolk Railway	MC 1959
51505	DMCL	Ecclesbourne Valley Railway	MC 1959
51511	DMCL	North Yorkshire Moors Railway	MC 1959
51512	DMCL	Cambrian Railway Trust, Llynclys	MC 1959
51803	DMCL	Keighley & Worth Valley Railway	MC 1959
56055–54055	DTCL	Cambrian Railway Trust, Llynclys	MC 1957
56062–54062	DTCL	North Norfolk Railway	MC 1957
56342–54342–042222	DTCL	Midland Railway-Butterley	MC 1958
56343–54343	DTCL	East Kent Light Railway	MC 1958
56347–54347	DTCL	Bressingham Steam Museum	MC 1958
56352–54352	DTCL	East Lancashire Railway (N)	MC 1958
56356–54356–6300h	DTCL	Gloucestershire-Warwickshire Railway	MC 1959

56358–54358		DTCL	East Anglian Railway Museum	MC 1959
56365–54365		DTCL	East Anglian Railway Museum	MC 1958
56408		DTCL	Spa Valley Railway	MC 1958
59117		TCL	Mid Norfolk Railway	MC 1958
59303		TSL	Midland Railway-Butterley	MC 1957
59539		TCL	North Yorkshire Moors Railway	MC 1959

CLASS 103 — PARK ROYAL TWIN UNITS

Built: 1958. Normal formation: DMBS–DTCL.
Engines: Two AEC 220 of 112 kW (150 h.p.).

DMBS	18.49 x 2.82 m	34 tonnes	–/52
DTCL	18.49 x 2.82 m	27 tonnes	16/48

50397	DMBS	Amman Valley Railway	PR 1958
50413	DMBS	Helston Railway, Gwinear Road near Camborne	PR 1958
56160–DB 975228	DTCL	Denbigh & Mold Junction Railway, Sodom	PR 1958
56169	DTCL	Helston Railway, Gwinear Road near Camborne	PR 1958

Note: 50397 was allocated number DB 975137 but this was never carried.

CLASS 104 — BRCW UNITS

Built: 1957–58. Various formations.
Engines: Two BUT (Leyland) of 112 kW (150 h.p.).

DMBS	18.49 x 2.82 m	31.5 tonnes	–/52
TCL	18.49 x 2.82 m	24.5 tonnes	12/54
TBSL	18.49 x 2.82 m	25.5 tonnes	–/51
DMCL	18.49 x 2.82 m	31.5 tonnes	12/54 (12/51*)
DTCL	18.49 x 2.82 m	24.5 tonnes	12/54

50437–53437	DMBS	Churnet Valley Railway	BRCW 1957
50447–53447	DMBS	Llangollen Railway	BRCW 1957
50454–53454	DMBS	Llangollen Railway	BRCW 1957
50455–53455	DMBS	Churnet Valley Railway	BRCW 1957
50479–53479	DMBS	Telford Steam Railway	BRCW 1958
50494–53494	DMCL	Churnet Valley Railway	BRCW 1957
50517–53517	DMCL	Churnet Valley Railway	BRCW 1957
50528–53528	DMCL	Llangollen Railway	BRCW 1958
50531–53531	DMCL	Telford Steam Railway	BRCW 1958
50556–53556*	DMCL	Telford Steam Railway	BRCW 1958
56182–54182–977554	DTCL	Churnet Valley Railway	BRCW 1958
59137	TCL	Churnet Valley Railway	BRCW 1957
59228	TBSL	Telford Steam Railway	BRCW 1958

CLASS 105 — CRAVEN TWIN UNITS

Built: 1957–59. Normal formation: DMBS–DTCL or DMCL.
Engines: Two AEC 220 of 112 kW (150 h.p.).

DMBS	18.49 x 2.82 m	29.5 tonnes	–/52
DTCL	18.49 x 2.82 m	23.5 tonnes	12/51

51485	DMBS	East Lancashire Railway	Cravens 1959
56121	DTCL	East Lancashire Railway	Cravens 1957
56456–54456	DTCL	Llangollen Railway	Cravens 1959

CLASS 107   DERBY HEAVYWEIGHT TRIPLE UNITS
Built: 1960–61. Normal formation: DMBS–TSL–DMCL.
Engines: Two BUT (Leyland) of 112 kW (150 h.p.).
Max. Speed: 70 m.p.h.

DMBS	18.49 x 2.82 m	35 tonnes	–/52
DMCL	18.49 x 2.82 m	35.5 tonnes	12/53
TSL	18.49 x 2.82 m	28.5 tonnes	–/71

51990–977830	DMBS	Strathspey Railway	Derby 1960
51993–977834	DMBS	Embsay & Bolton Abbey Railway	Derby 1961
52005–977832	DMBS	Embsay & Bolton Abbey Railway	Derby 1961
52006	DMBS	Embsay & Bolton Abbey Railway	Derby 1961
52008	DMBS	Strathspey Railway	Derby 1961
52012–977835	DMCL	Embsay & Bolton Abbey Railway	Derby 1960
52025–977833	DMCL	Embsay & Bolton Abbey Railway	Derby 1961
52029 h	DMCL	Llanelli & Mynydd Mawr Railway, Cynheidre	Derby 1961
52030–977831	DMBS	Strathspey Railway	Derby 1961
52031	DMCL	Embsay & Bolton Abbey Railway	Derby 1961
59791	TSL	Embsay & Bolton Abbey Railway	Derby 1961

Note: 52029 is stored at MoD Llangennech.

CLASS 108   DERBY LIGHTWEIGHT UNITS
Built: 1958–1961. Various formations.
Engines: Two Leyland of 112 kW (150 h.p.).

DMBS	18.49 x 2.79 m	29.5 tonnes	–/52
TBSL	18.49 x 2.79 m	21.5 tonnes	–/50
TSL	18.49 x 2.79 m	21.5 tonnes	–/68
DMCL	18.49 x 2.79 m	28.5 tonnes	12/53
DTCL	18.49 x 2.79 m	21.5 tonnes	12/53

50599–53599	DMBS	East Anglian Railway Museum	Derby 1958
50619–53619	DMBS	Dean Forest Railway	Derby 1958
50627–53627–977853	DMBS	Peak Railway	Derby 1958
50628–53628	DMBS	Keith & Dufftown Railway	Derby 1958
50632–53632	DMCL	Pontypool & Blaenavon Railway	Derby 1958
50645–53645	DMCL	Nottingham Heritage Centre, Ruddington	Derby 1958
50926–53926–977814	DMBS	Nottingham Heritage Centre, Ruddington	Derby 1959
50928–53928	DMBS	Keighley & Worth Valley Railway	Derby 1959
50933–53933	DMBS	Peak Railway	Derby 1960
50971–53971	DMBS	Kent & East Sussex Railway	Derby 1959
50980–53980	DMBS	Bodmin Steam Railway	Derby 1959
51562	DMCL	National Railway Museum	Derby 1959
51565	DMCL	Keighley & Worth Valley Railway	Derby 1959
51566	DMCL	Peak Railway	Derby 1959
51567-977854	DMCL	Peak Railway	Derby 1959
51568	DMCL	Keith & Dufftown Railway	Derby 1959
51571	DMCL	Kent & East Sussex Railway	Derby 1960
51572	DMCL	Wensleydale Railway	Derby 1960
51907	DMBS	Llangollen Railway	Derby 1960
51909	DMBS	Avon Valley Railway	Derby 1960
51914	DMBS	Dean Forest Railway	Derby 1960
51919	DMBS	Barry Island Railway	Derby 1960
51922	DMBS	National Railway Museum	Derby 1960
51933	DMBS	Swanage Railway	Derby 1960
51935	DMBS	Severn Valley Railway	Derby 1960
51937–977806	DMBS	Peak Railway	Derby 1960
51941	DMBS	Severn Valley Railway	Derby 1960
51942	DMBS	Pontypool & Blaenavon Railway	Derby 1961
51947	DMBS	Bodmin Steam Railway	Derby 1961
51950	DMBS	Gloucestershire-Warwickshire Railway	Derby 1961
52044	DMCL	Pontypool & Blaenavon Railway	Derby 1960

52048	DMCL	Barry Island Railway	Derby	1960
52053–977807	DMCL	Keith & Dufftown Railway	Derby	1960
52054	DMCL	Bodmin Steam Railway	Derby	1960
52062	DMCL	Gloucestershire-Warwickshire Railway	Derby	1961
52064	DMCL	Severn Valley Railway	Derby	1961
56207–54207 h	DTCL	Appleby-Frodingham RPS	Derby	1958
56208–54208	DTCL	Severn Valley Railway	Derby	1958
56223–54223	DTCL	East Anglian Railway Museum	Derby	1959
56224–54224	DTCL	Ecclesbourne Valley Railway	Derby	1959
56270–54270	DTCL	Pontypool & Blaenavon Railway	Derby	1959
56271–54271	DTCL	Avon Valley Railway	Derby	1960
56274–54274 h	DTCL	Rutland Railway Museum	Derby	1960
56279–54279	DTCL	Barry Island Railway	Derby	1960
56484–54484	DTCL	Peak Railway	Derby	1960
56490–54490	DTCL	Llangollen Railway	Derby	1960
56491–54491	DTCL	Keith & Dufftown Railway	Derby	1960
56492–54492	DTCL	Dean Forest Railway	Derby	1960
56495–54495	DTCL	Dean Forest Railway	Derby	1960
56504–54504	DTCL	Swanage Railway	Derby	1960
59245 h	TBSL	Appleby-Frodingham RPS	Derby	1958
59250 "THE CHATHAM BAR"	TBSL	Severn Valley Railway	Derby	1958
59387	TSL	Peak Railway	Derby	1958

Notes: 51909 and 56271 are currently under restoration at Long Marston Workshops.
50628 and 56491 are named "SPIRIT OF DUFFTOWN" and 51568 & 52053 are named "SPIRIT OF BANFFSHIRE".

CLASS 109   D. WICKHAM TWIN UNITS
Built: 1957. Normal formation: DMBS–DTCL.
Engines: Two BUT (Leyland) of 112 kW (150 h.p.).

DMBS	18.49 x 2.82 m	27.5 tonnes	–/52	
DTCL	18.49 x 2.82 m	20.5 tonnes	16/50	

50416–DB 975005	DMBS	Llangollen Railway	Wkm 1957
56171–DB 975006	DTCL	Llangollen Railway	Wkm 1957

CLASS 110   BRCW CALDER VALLEY UNITS
Built: 1961–2. Normal formation: DMBC–TSL–DMCL.
Engines: Two Rolls-Royce C6NFLH38D of 134 kW (180 h.p.).

DMBC	18.48 x 2.82 m	32.5 tonnes	12/33	
DMCL	18.48 x 2.82 m	32.5 tonnes	12/54	
TSL	18.48 x 2.82 m	24.5 tonnes	–/72	

51813	DMBC	Wensleydale Railway	BRCW 1961
51842	DMCL	Wensleydale Railway	BRCW 1961
52071	DMBC	Lakeside & Haverthwaite Railway	BRCW 1962
52077	DMCL	Lakeside & Haverthwaite Railway	BRCW 1961
59701	TSL	Wensleydale Railway	BRCW 1961

CLASS 111   METRO-CAMMELL TRAILER BUFFET
Built: 1960. Used to augment other units as required.

TSLRB	18.49 x 2.82 m	25.5 tonnes	–/53	

59575	TSLRB	Great Central Railway	MC 1960

135

CLASS 114   DERBY HEAVYWEIGHT TWIN UNITS
Built: 1956–57. Normal formation: DMBS–DTCL.
Engines: Two Leyland Albion of 149 kW (200 h.p.).

DMBS	20.45 x 2.82 m	38 tonnes	–/62
DTCL	20.45 x 2.82 m	30 tonnes	12/62

50015–53015–55929–977775	DMBS	Midland Railway-Butterley		Derby 1957
50019–53019	DMBS	Midland Railway-Butterley		Derby 1957
56006–54006	DTCL	Midland Railway-Butterley		Derby 1956
56015–54015–54904–977776	DTCL	Midland Railway-Butterley		Derby 1957
56047–54047	DTCL	Strathspey Railway		Derby 1957

CLASS 115   DERBY SUBURBAN QUAD UNITS
Built: 1960. Non-gangwayed when built, but gangways subsequently fitted. Normal formation: DMBS–TSso–TCL–DMBS.
Engines: Two Leyland Albion of 149 kW (200 h.p.).

DMBS	20.45 x 2.82 m	38.5 tonnes	–/74 (originally –/78)
TCL	20.45 x 2.82 m	30.5 tonnes	28/38 (originally 30/40)
TSso	20.45 x 2.82 m	29.5 tonnes	–/98 (originally –/106)

BR	Present			
51655		DMBS	Barry Island Railway	Derby 1960
51669		DMBS	Spa Valley Railway	Derby 1960
51677		DMBS	Barry Island Railway	Derby 1960
51849		DMBS	Spa Valley Railway	Derby 1960
51852		DMBS	West Somerset Railway	Derby 1960
51859		DMBS	West Somerset Railway	Derby 1960
51880		DMBS	West Somerset Railway	Derby 1960
51886		DMBS	Buckinghamshire Railway Centre	Derby 1960
51887		DMBS	West Somerset Railway	Derby 1960
51899		DMBS	Buckinghamshire Railway Centre	Derby 1960
59659 h	"9659"	TSso	South Devon Railway	Derby 1960
59664		TCL	Barry Island Railway	Derby 1960
59678		TCL	West Somerset Railway	Derby 1960
59719		TCL	South Devon Railway	Derby 1960
59740 h	"9740"	TSso	South Devon Railway	Derby 1960
59761		TCL	Buckinghamshire Railway Centre	Derby 1960

Note: 51899 is named "AYLESBURY COLLEGE SILVER JUBILEE"

CLASS 116   DERBY SUBURBAN TRIPLE UNITS
Built: 1957–58. Non-gangwayed when built, but gangways subsequently fitted. Normal formation: DMBS–TS or TC–DMS.
Engines: Two Leyland of 112 kW (150 h.p.).

DMBS	20.45 x 2.82 m	36.5 tonnes	–/65
DMS	20.45 x 2.82 m	36.5 tonnes	–/89 (originally –/95)
TS§	20.45 x 2.82 m	29 tonnes	–/98 (originally –/102)
TC	20.45 x 2.82 m	29 tonnes	20/68 (originally 28/74)

§-converted from TC seating 28/74.

51131	DMBS	Battlefield Railway	Derby 1958
51134	DMBS	Swansea Vale Railway	Derby 1958
51135	DMBS	Swansea Vale Railway	Derby 1958
51138–977921	DMBS	Nottingham Heritage Centre, Ruddington	Derby 1958
51147	DMS	Swansea Vale Railway	Derby 1958
51148	DMS	Swansea Vale Railway	Derby 1958
51151	DMS	Nottingham Heritage Centre, Ruddington	Derby 1958
59003 h	TS	Paignton & Dartmouth Railway	Derby 1957
59004 h "Emma"	TS	Paignton & Dartmouth Railway	Derby 1957
59444 h	TC	Chasewater Light Railway	Derby 1958
59445	TC	Swansea Vale Railway	Derby 1958

CLASS 117 PRESSED STEEL SUBURBAN TRIPLE UNITS

Built: 1960. Non-gangwayed when built, but gangways subsequently fitted. Normal formation: DMBS–TCL–DMS.
Engines: Two Leyland of 112 kW (150 h.p.).

DMBS	20.45 x 2.82 m		36.5 tonnes	–/65
TCL	20.45 x 2.82 m		30.5 tonnes	22/48 (originally 24/50)
DMS	20.45 x 2.82 m		36.5 tonnes	–/89 (originally –/91)

51339	DMBS	Barry Island Railway	PS 1960
51341	DMBS	Midland Railway-Butterley	PS 1960
51342	DMBS	Epping & Ongar Railway	PS 1960
51346	DMBS	Swanage Railway	PS 1960
51347	DMBS	Nene Valley Railway	PS 1960
51351	DMBS	Pontypool & Blaenavon Railway	PS 1960
51352	DMBS	St. Modwen Properties, Long Marston	PS 1960
51353	DMBS	Midland Railway-Butterley	PS 1960
51354	DMBS	Llanelli & Mynydd Mawr Railway, Cynheidre	PS 1960
51356	DMBS	Weardale Railway	PS 1960
51359	DMBS	Northampton & Lamport Railway	PS 1960
51360	DMBS	Ecclesbourne ValleyRailway	PS 1960
51363	DMBS	Mid Hants Railway	PS 1960
51365	DMBS	Plym Valley Railway	PS 1960
51367	DMBS	Strathspey Railway	PS 1960
51370	DMBS	Titley Junction Station, Herefordshire	PS 1960
51372	DMBS	Chasewater Light Railway	PS 1960
51376	DMS	St. Modwen Properties, Long Marston	PS 1960
51381	DMS	Mangapps Farm, Railway Museum	PS 1960
51382	DMS	Barry Island Railway	PS 1960
51384	DMS	Epping & Ongar Railway	PS 1960
51388	DMS	Swanage Railway	PS 1960
51392	DMS	Weardale Railway	PS 1960
51395	DMS	Midland Railway-Butterley	PS 1960
51396	DMS	Llanelli & Mynydd Mawr Railway, Cynheidre	PS 1960
51397	DMS	Pontypool & Blaenavon Railway	PS 1960
51398	DMS	Midland Railway-Butterley	PS 1960
51400	DMS	Mid Hants Railway	PS 1960
51401	DMS	Nene Valley Railway	PS 1960
51402	DMS	Strathspey Railway	PS 1960
51405	DMS	Mid Hants Railway	PS 1960
51407	DMS	Plym Valley Railway	PS 1960
51412	DMS	Titley Junction Station, Herefordshire	PS 1960
59486	TCL	Midland Railway-Butterley	PS 1960
59488 h	TCL	Paignton & Dartmouth Railway	PS 1960
59490	TCL	Swansea Vale Railway	PS 1960
59492	TCL	Weardale Railway	PS 1960
59493 h	TCL	West Somerset Railway	PS 1960
59494 h "CHLOE"	TCL	Paignton & Dartmouth Railway	PS 1960
59496 h	TCL	Ecclesbourne Valley Railway	PS 1960
59500	TCL	Wensleydale Railway	PS 1960
59501	TCL	Nottingham Heritage Centre, Ruddington	PS 1960
59503 h	TCL	Paignton & Dartmouth Railway	PS 1960
59505	TCL	St. Modwen Properties, Long Marston	PS 1960
59506	TCL	West Somerset Railway	PS 1960
59507 h	TCL	Paignton & Dartmouth Railway	PS 1960
59508	TCL	Nene Valley Railway	PS 1960
59509	TCL	Wensleydale Railway	PS 1960
59510	TCL	Mid Hants Railway	PS 1960
59511	TCL	Strathspey Railway	PS 1960
59513 h	TCL	Paignton & Dartmouth Railway	PS 1960
59514	TCL	Swindon & Cricklade Railway	PS 1960
59515 h	TCL	West Somerset Railway	PS 1960
59516	TCL	Swanage Railway	PS 1960

59517 h	TCL	Paignton & Dartmouth Railway	PS 1960
59520	TCL	Pontypool & Blaenavon Railway	PS 1960
59521	TCL	Midland Railway-Butterley	PS 1960
59522 h	TCL	Chasewater Light Railway	PS 1960

Note: 51354 and 51396 are stored at MoD Llangennech.

CLASS 118 — BRCW SUBURBAN TRIPLE UNITS

Built: 1960. Non-gangwayed when built, but gangways subsequently fitted. Normal formation: DMBS–TCL–DMS.
Engines: Two Leyland of 112 kW (150 h.p.).

DMS	20.45 x 2.82 m	36.5 tonnes	–/89 (originally –/91)

51321–977753	DMS	Battlefield Railway	BRCW 1960

CLASS 119 — GRCW CROSS-COUNTRY UNITS

Built: 1959. Normal formation: DMBC–TSLRB–DMSL.
Engines: Two Leyland of 112 kW (150 h.p.).

DMBC	20.45 x 2.82 m	37.5 tonnes	18/16
DMSL	20.45 x 2.82 m	38.5 tonnes	–/68

51073	DMBC	Midland Railway-Butterley	GRCW 1959
51074	DMBC	Swindon & Cricklade Railway	GRCW 1959
51104	DMSL	Swindon & Cricklade Railway	GRCW 1959

CLASS 120 — SWINDON CROSS-COUNTRY UNITS

Built: 1958. Normal formation: DMBC–TSLRB–DMSL.

TSLRB	20.45 x 2.82 m	31.5 tonnes	–/60

59276	TSLRB	Great Central Railway	Swindon 1958

CLASS 121 — PRESSED STEEL SINGLE UNITS & DRIVING TRAILERS

Built: 1960–61. Non-gangwayed single cars with two driving cabs plus driving trailers used for augmentation. The driving trailers were latterly fitted with gangways for coupling to power cars of other classes.
Engines: Two Leyland of 112 kW (150 h.p.).
Transmission: Mechanical.

DMBS	20.45 x 2.82 m	38 tonnes	–/65
DTS	20.45 x 2.82 m	30 tonnes	–/89 (originally –/91)

55023	DMBS	Chinnor & Princes Risborough Railway	PS 1960
55026–977824	DMBS	Swansea Vale Railway	PS 1960
55033–977826	DMBS	Colne Valley Railway	PS 1960
55034–977828	DMBS	Tyseley Locomotive Works, Birmingham	PS 1961
56285–54285–977486 h	DTS	Northamptonshire Ironstone Railway	PS 1961
56287–54287 h	DTS	Colne Valley Railway	PS 1961
56289–54289 h	DTS	Ecclesbourne Valley Railway	PS 1961

CLASS 122 — GRCW SINGLE UNITS

Built: 1958. Non-gangwayed single cars with two driving cabs.
Engines: Two AEC 220 of 112 kW (150 h.p.).

DMBS	20.45 x 2.82 m	36.5 tonnes	–/65

55000	DMBS	South Devon Railway	GRCW 1958
55001-DB975023	DMBS	Northampton & Lamport Railway	GRCW 1958
55003	DMBS	Mid Hants Railway	GRCW 1958
55005	DMBS	Battlefield Railway	GRCW 1958
55006	DMBS	Ecclesbourne Valley Railway	GRCW 1958
55009	DMBS	Mid Norfolk Railway	GRCW 1958

CLASS 126 — SWINDON INTER-CITY UNITS
Built: 1956–59. Various formations.
Engines: Two AEC 220 of 112 kW (150 h.p.).

DMBSL	20.45 x 2.82 m	38.5 tonnes	–/52	
TCK	20.45 x 2.82 m	32.3 tonnes	18/32	
TFKRB	20.45 x 2.82 m	34 tonnes	18/12	
DMSL	20.45 x 2.82 m	38.5 tonnes	–/64	

51017	DMSL	Bo'ness & Kinneil Railway	Swindon 1959
51043	DMBSL	Bo'ness & Kinneil Railway	Swindon 1959
59404	TCK	Bo'ness & Kinneil Railway	Swindon 1959
79443	TFKRB	Bo'ness & Kinneil Railway	Swindon 1957

Note: 59404 converted by BR to TSK–/56 but restored as TCK

CLASS 127 — DERBY SUBURBAN QUAD UNITS
Built: 1959. Non-gangwayed. Normal formation: DMBS–TSL–TS–DMBS. Some DMBS rebuilt as DMPMV Normal formation: DMPMV(A)-DMPMV(B).
Engines: Two Rolls-Royce C8 of 177 kW (238 h.p.).
Transmission: Hydraulic.

DMBS	20.45 x 2.82 m	40.6 tonnes	–/76	
DMPMV (A)	20.45 x 2.82 m	40 tonnes		
DMPMV (B)	20.45 x 2.82 m	40 tonnes		
TSL	20.45 x 2.82 m	30.5 tonnes	–/86	

51592	DMBS	South Devon Railway	Derby 1959
51604	DMBS	South Devon Railway	Derby 1959
51616	DMBS	Great Central Railway	Derby 1959
51618	DMBS	Llangollen Railway	Derby 1959
51622	DMBS	Great Central Railway	Derby 1959
51591–55966	DMPMV(A)	Midland Railway-Butterley	Derby 1959
51610–55967 "Glen Ord"	DMPMV(B)	Rogart Station	Derby 1959
51625–55976	DMPMV(A)	Midland Railway-Butterley	Derby 1959
59603	TSL	Chasewater Light Railway	Derby 1959
59609	TSL	Midland Railway-Butterley	Derby 1959

UNCLASSIFIED — DERBY LIGHTWEIGHT TWIN UNIT
Built: 1955. Normal formation. DMBS–DTCL.
Engines: Two BUT (AEC) of 112 kW (150 h.p.).

DMBS	18.49 x 2.82 m	27.4 tonnes	–/61	
DTCL	18.49 x 2.82 m	21.3 tonnes	9/53	

79018–DB 975007	DMBS	Midland Railway-Butterley	Derby 1955
79612–DB 975008	DTCL	Midland Railway-Butterley	Derby 1955

UNCLASSIFIED — DERBY LIGHTWEIGHT SINGLE UNIT
Built: 1956. Non-gangwayed single cars with two driving cabs.
Engines: Two AEC of 112 kW (150 h.p.).

DMBS	18.49 x 2.82 m	27 tonnes	–/52	

79900–DB975010 "IRIS"	DMBS	Midland Railway-Butterley	Derby 1956

5.2.3. FOUR-WHEELED DIESEL RAILBUSES
UNCLASSIFIED　　　WAGGON UND MASCHINENBAU
Built: 1958. 5 built.
Engine: Buessing of 112 kW (150 h.p.) at 1900 rpm. (§ AEC 220 of 112 kW (150h.p.))
Max. Speed: 70 m.p.h.

DMS　　　13.95 x 2.67 m　　　15 tonnes　　　–/56

79960		North Norfolk Railway	WMD 1265/1958
79962		Keighley & Worth Valley Railway	WMD 1267/1958
79963		Mangapps Farm, Burnham-on-Crouch	WMD 1268/1958
79964§		Keighley & Worth Valley Railway	WMD 1298/1958

UNCLASSIFIED　　　AC CARS
Built: 1958. 5 built.
Engine: AEC 220 of 112 kW (150 h.p.). (§ engine removed)
Max. Speed: 70 m.p.h.

DMS　　　11.33 x 2.82 m　　　11 tonnes　　　–/46

79976§		Great Central Railway	AC 1958
79978		Colne Valley Railway	AC 1958

UNCLASSIFIED　　　BR DERBY/LEYLAND
Built: 1977.
Engine: Leyland 510 of 149 kW (200 h.p.). (Fitted 1979).
Transmission: Mechanical. Self Changing Gears.
Max. Speed: 75 m.p.h.　　　Air braked.

DMS　　　12.32 x 2.50 m　　　16.67 tonnes　　　–/40

R1-RDB 975874　　　North Norfolk Railway (N)　　　RTC Derby 1977

UNCLASSIFIED　　　BREL DERBY/LEYLAND
Built: 1981.
Engine: Leyland 690 of 149 kW (200 h.p.).
Transmission: Mechanical. Self Changing Gears SE4 epicyclic gearbox and cardan shafts to SCG RF28 final drive.
Max. Speed: 75 m.p.h.　　　Air braked.
Gauge: Built as 1435 mm but converted to 1600 mm when sold to Northern Ireland Railways.

DMS　　　15.30 x 2.50 m　　　19.96 tonnes　　　–/56

R3.03–RDB 977020　　RB3　　Downpatrick Steam Railway, NI　　RTC Derby 1981

CLASS 140
DERBY/LEYLAND BUS PROTOTYPE TWIN RAILBUS
Built: 1981.
Engine: Leyland TL11 of 152 kW (205 h.p.).
Transmission: Mechanical. Self-Changing Gears 4-speed gearbox.
Max. Speed: 75 m.p.h.　　　Air braked.

DMSL　　　16.20 x 2.50 m　　　23.2 tonnes　　　–/50
DMS　　　16.20 x 2.50 m　　　23.0 tonnes　　　–/52

55500	DMS	Keith & Dufftown Railway	Derby 1981
55501	DMSL	Keith & Dufftown Railway	Derby 1981

CLASS 141 BREL/LEYLAND BUS TWIN RAILBUS

Built: 1983–84. Modified by Barclay 1988–89.
Engine: Leyland TL11 of 152 kW (205 h.p.)
Transmission: Hydraulic. Voith T211r.
Max. Speed: 75 m.p.h.   Air braked.

DMS	15.45 x 2.50 m	26.0 tonnes	–/50
DMSL	15.45 x 2.50 m	26.5 tonnes	–/44

55502	DMS	Iranian Islamic Republic Railways	BRE-Leyland 1983
55503	DMS	Weardale Railway	BRE-Leyland 1984
55505	DMS	Iranian Islamic Republic Railways	BRE-Leyland 1984
55506	DMS	Connexion, Ultrecht, Netherlands	BRE-Leyland 1984
55507	DMS	Iranian Islamic Republic Railways	BRE-Leyland 1984
55508	DMS	Colne Valley Railway	BRE-Leyland 1984
55509	DMS	Iranian Islamic Republic Railways	BRE-Leyland 1984
55510	DMS	Weardale Railway	BRE-Leyland 1984
55511	DMS	Iranian Islamic Republic Railways	BRE-Leyland 1984
55512	DMS	Connexion, Ultrecht, Netherlands	BRE-Leyland 1984
55513	DMS	Midland Railway-Butterley	BRE-Leyland 1984
55514	DMS	Iranian Islamic Republic Railways	BRE-Leyland 1984
55515	DMS	Iranian Islamic Republic Railways	BRE-Leyland 1984
55516	DMS	Iranian Islamic Republic Railways	BRE-Leyland 1984
55517	DMS	Iranian Islamic Republic Railways	BRE-Leyland 1984
55518	DMS	Iranian Islamic Republic Railways	BRE-Leyland 1984
55519	DMS	Iranian Islamic Republic Railways	BRE-Leyland 1984
55520	DMS	Iranian Islamic Republic Railways	BRE-Leyland 1984
55522	DMSL	Iranian Islamic Republic Railways	BRE-Leyland 1983
55523	DMSL	Weardale Railway	BRE-Leyland 1984
55525	DMSL	Iranian Islamic Republic Railways	BRE-Leyland 1984
55526	DMSL	Connexion, Utrecht, Netherlands	BRE-Leyland 1984
55527	DMSL	Iranian Islamic Republic Railways	BRE-Leyland 1984
55528	DMSL	Colne Valley Railway	BRE-Leyland 1984
55529	DMSL	Iranian Islamic Republic Railways	BRE-Leyland 1984
55530	DMSL	Weardale Railway	BRE-Leyland 1984
55531	DMSL	Iranian Islamic Republic Railways	BRE-Leyland 1984
55532	DMSL	Connexion, Ultrecht, Netherlands	BRE-Leyland 1984
55533	DMS	Midland Railway-Butterley	BRE-Leyland 1984
55534	DMSL	Iranian Islamic Republic Railways	BRE-Leyland 1984
55535	DMSL	Iranian Islamic Republic Railways	BRE-Leyland 1984
55536	DMSL	Iranian Islamic Republic Railways	BRE-Leyland 1984
55537	DMSL	Iranian Islamic Republic Railways	BRE-Leyland 1984
55538	DMSL	Iranian Islamic Republic Railways	BRE-Leyland 1984
55539	DMSL	Iranian Islamic Republic Railways	BRE-Leyland 1984
55540	DMSL	Iranian Islamic Republic Railways	BRE-Leyland 1984

5.2.4. DIESEL ELECTRIC MULTIPLE UNITS

CLASS 201 "HASTINGS" 6-CAR DIESEL-ELECTRIC UNIT
Built: 1957 by BR Eastleigh Works on frames constructed at Ashford. Special narrow-bodied units built to the former loading gauge of the Tonbridge–Battle line.
Normal formation: DMBSO–TSOL–TSOL–TFK–TSOL–DMBSO.
Engines: English Electric 4SRKT of 370 kW (500 h.p.).
Transmission: Two EE 507 traction motors on the inner power car bogie.
Max. Speed: 75 m.p.h.

DMBSO	18.35 x 2.50 m	54 tonnes	–/22
TSOL	18.35 x 2.50 m	29 tonnes	–/52
TFK	18.36 x 2.50 m	30 tonnes	42/–

60000	"Hastings"	DMBSO	St. Leonards Railway Engineering	Eastleigh 1957
60001		DMBSO	St. Leonards Railway Engineering	Eastleigh 1957
60500		TSOL	St. Leonards Railway Engineering	Eastleigh 1957
60501		TSOL	St. Leonards Railway Engineering	Eastleigh 1957
60502		TSOL	St. Leonards Railway Engineering	Eastleigh 1957
60700		TFK	St. Leonards Railway Engineering	Eastleigh 1957

CLASS 202 "HASTINGS" 6-CAR DIESEL-ELECTRIC UNITS
Built: 1957–58 by BR Eastleigh Works on frames constructed at Ashford. Special narrow-bodied units built to the former loading gauge of the Tonbridge–Battle line.
Normal formation: DMBSO–TSOL–TSOL (or TRB)–TFK–TSOL–DMBSO.
Engines: English Electric 4SRKT of 370 kW (500 h.p.).
Transmission: Two EE 507 traction motors on the inner power car bogie.
Max. Speed: 75 m.p.h.

DMBSO	20.34 x 2.50 m	55 tonnes	–/30
TSOL	20.34 x 2.50 m	29 tonnes	–/60
TFK	20.34 x 2.50 m	31 tonnes	48/–
TRB	20.34 x 2.50 m	34 tonnes	–/21

60016	"60116""Mountfield"	DMBSO	St. Leonards Railway Engineering	Eastleigh 1957
60018	"60118""Tunbridge Wells"	DMBSO	St. Leonards Railway Engineering	Eastleigh 1957
60019		DMBSO	St. Leonards Railway Engineering	Eastleigh 1957
60527		TSOL	St. Leonards Railway Engineering	Eastleigh 1957
60528		TSOL	St. Leonards Railway Engineering	Eastleigh 1957
60529		TSOL	St. Leonards Railway Engineering	Eastleigh 1957
60708		TFK	St. Leonards Railway Engineering	Eastleigh 1957
60709		TFK	St. Leonards Railway Engineering	Eastleigh 1957
60750–RDB 975386		TRB	The Pump House Steam & Transport Museum	Eastleigh 1958

CLASS 205 "HAMPSHIRE" 3-CAR DIESEL-ELECTRIC UNITS
Built: 1958–59 by BR Eastleigh Works on frames constructed at Ashford. Non-gangwayed.
Normal formation: DMBSO–TSO–DTCsoL.
Engines: English Electric 4SRKT of 370 kW (500 h.p.).
Transmission: Two EE 507 traction motors on the inner power car bogie.
Max. Speed: 75 m.p.h.

DMBSO	20.34 x 2.82 m	56 tonnes	–/52
TSO	20.28 x 2.82 m	30 tonnes	–/104
DTCsoL	20.34 x 2.82 m	32 tonnes	19/50

60100–60154	DMBSO	East Kent Light Railway	Eastleigh 1957
60108	DMBSO	Eden Valley Railway, Warcop	Eastleigh 1957
60110	DMBSO	North Yorkshire Moors Railway	Eastleigh 1957
60117	DMBSO	Pontypool & Blaenavon Railway	Eastleigh 1957
60122	DMBSO	Lavender Line, Isfield	Eastleigh 1959
60124	DMBSO	Mid Hants Railway	Eastleigh 1959
60146	DMBSO	Dartmoor Railway, Okehampton	Eastleigh 1962
60150	DMBSO	Dartmoor Railway, Okehampton	Eastleigh 1962
60151	DMBSO	Lavender Line, Isfield	Eastleigh 1962

60658	TSO	Eden Valley Railway, Warcop	Eastleigh 1959
60669	TSO	Marshalls Transport, Pershore Airfield	Eastleigh 1959
60673	TSO	Dartmoor Railway, Okehampton	Eastleigh 1962
60677	TSO	Dartmoor Railway, Okehampton	Eastleigh 1962
60678	TSO	Lavender Line, Isfield	Eastleigh 1962
60800	DTCsoL	East Kent Light Railway	Eastleigh 1957
60808	DTCsoL	Eden Valley Railway, Warcop	Eastleigh 1957
60810	DTCsoL	North Yorkshire Moors Railway	Eastleigh 1957
60811	DTCsoL	St. Leonards Railway Engineering	Eastleigh 1957
60820	DTCsoL	St. Leonards Railway Engineering	Eastleigh 1958
60822	DTCsoL	Allely's, The Slough, Studley	Eastleigh 1959
60824	DTCsoL	Mid Hants Railway	Eastleigh 1959
60827	DTCsoL	Dartmoor Railway, Okehampton	Eastleigh 1962
60828	DTCsoL	Pontypool & Blaenavon Railway	Eastleigh 1962
60831	DTCsoL	Dartmoor Railway, Okehampton	Eastleigh 1962
60832	DTCsoL	Lavender Line, Isfield	Eastleigh 1962

CLASS 207   "OXTED" 3-CAR DIESEL-ELECTRIC UNITS

Built: 1962 by BR Eastleigh Works on frames constructed at Ashford. Reduced body width to allow operation through Somerhill Tunnel. Non-gangwayed.
Normal formation: DMBSO–TCsoL–DTSO.
Engines: English Electric 4SRKT of 370 kW (500 h.p.).
Transmission: Two EE 507 traction motors on the inner power car bogie.
Max. Speed: 75 m.p.h.

DMBSO	20.34 x 2.74 m	56 tonnes	–/42
DTSO	20.32 x 2.74 m	32 tonnes	–/76
TCsoL	20.34 x 2.74 m	31 tonnes	24/42

60127	DMBSO	Swindon & Crickdale Railway	Eastleigh 1962
60130	DMBSO	East Lancashire Railway	Eastleigh 1962
60138–977907	DMBSO	The Pump House Steam & Transport Museum	Eastleigh 1962
60142	DMBSO	Spa Valley Railway	Eastleigh 1962
60616	TCsoL	East Somerset Railway	Eastleigh 1962
60901	DTSO	Swindon & Cricklade Railway	Eastleigh 1962
60904	DTSO	East Lancashire Railway	Eastleigh 1962
60916	DTSO	Spa Valley Railway	Eastleigh 1962

CLASS 210   DERBY PROTOTYPE
3-CAR (*4-CAR) DIESEL-ELECTRIC UNITS

Built: 1981 by BR Derby Works.
Normal formation: DMSO–TSO–DTSO (*DMBSO–TSO–TSOL–DTSO).
Engines: Ruston-Paxman 6RP200 of 840 kW (*MTU 12V396TC11 of 850 kW).
Transmission:
Max. Speed: 75 m.p.h.

DTSO	20.52 x 2.82 m	29 tonnes	–/74

54000–60300–67300*	DTSO	Coventry Railway Centre	Derby 1981
54001–60301–67301	DTSO	Allely's, The Slough, Studley	Derby 1981

6. ELECTRIC MULTIPLE UNITS

Type Codes
For type codes see the DMU section (section 5).

6.1. SOUTHERN RAILWAY EMU STOCK

CLASS 487 — WATERLOO & CITY LINE UNITS
Built: 1940. No permanent formations.
System: 630 V DC third rail.
Traction Motors: Two EE 500 of 140 kW (185 h.p.). **Max. Speed:** 35 m.p.h.

DMBTO	14.33 x 2.64 m	29 tons	–/40

BR	SR			
S 61 S	61	DMBTO	London Transport Depot Museum, Acton (N)	EE 1940

1285 CLASS (later 3 Sub) — SUBURBAN UNITS
Built: 1925. Normal formation. DMBT–TT–DMBT.
System: 630 V DC third rail.
Traction Motors: Two MV 167 kW (225 h.p.). **Max. Speed:** 75 m.p.h.

DMBT	18.90 x 2.44 m	39 tons	–/70

BR	SR				
S 8143 S	8143	DMBT	(ex-unit 1293 later 4308)	National Railway Museum	MC 1925

4 Cor "NELSONS" PORTSMOUTH EXPRESS STOCK
Built: 1937–38. Normal formation. DMBTO–TTK–TCK–DMBTO.
System: 630 V DC third rail.
Traction Motors: Two MV 167 kW (225 h.p.) per power car. **Max. Speed:** 75 m.p.h.

DMBTO	19.54 x 2.88 m	46.5 tons	–/52
TTK	19.54 x 2.85 m	32.65 tons	–/68
TCK	19.54 x 2.85 m	32.6 tons	30/24

BR	SR				
S 10096 S	10096	TTK	(ex-unit 3142)	East Kent Light Railway	Eastleigh 1937
S 11161 S	11161	DMBTO	(ex-unit 3142)	East Kent Light Railway	Eastleigh 1937
S 11179 S	11179	DMBTO	(ex-unit 3131)	National Railway Museum	Eastleigh 1937
S 11187 S	11187	DMBTO	(ex-unit 3135)	London Transport Depot Museum, Acton	Eastleigh 1937
S 11201 S	11201	DMBTO	(ex-unit 3142)	Bluebell Railway	Eastleigh 1937
S 11825 S	11825	TCK	(ex-unit 3142)	East Kent Light Railway	Eastleigh 1937

Note: S 11161 S was originally in unit 3065 and S11825S in unit 3135.

4 Sub (later Class 405) — SUBURBAN UNITS
Built: 1941–51. Normal formation DMBTO–TT–TTO–DMBTO.
System: 630 V DC third rail.
Traction Motors: Two EE507 of 185 kW (250 h.p.). **Max. Speed:** 75 m.p.h.

DMBTO	19.05 x 2.82 m	42 tons	–/82
TT	18.90 x 2.82 m	27 tons	–/120
TTO	18.90 x 2.82 m	26 tons	–/102

BR	SR				
S 10239 S	10239	TT	(ex-unit 4732)	Coventry Railway Centre	Eastleigh 1947
S 12354 S		TTO	(ex-unit 4732)	Coventry Railway Centre	Eastleigh 1948
S 12795 S		DMBTO	(ex-unit 4732)	Coventry Railway Centre	Eastleigh 1951
S 12796 S		DMBTO	(ex-unit 4732)	Coventry Railway Centre	Eastleigh 1951

Note: S 10239 S was originally in unit 4413 and S 12354 S in unit 4381.

2 Bil
Built: 1937. Normal formation DMBTK–DTCK.
System: 630 V DC third rail.
Traction Motors: Two EE of 205 kW (275 h.p.).

SEMI-FAST UNITS

Max. Speed: 75 m.p.h.

| DMBTK | 19.24 x 2.85 m | 43.5 tons | –/52 |
| DTCK | 19.24 x 2.85 m | 31.25 tons | 24/30 |

BR   *SR*
S 10656 S 10656 DMBTK (ex-unit 1890–2090) Locomotion: The NRM at Shildon Eastleigh 1937
S 12123 S 12123 DTCK   (ex-unit 1890–2090) Locomotion: The NRM at Shildon Eastleigh 1937

4 Buf
PORTSMOUTH EXPRESS STOCK
Built: 1937. Normal formation. DMBTO–TCK–TRBT–DMBTO.
System: 630 V DC third rail.   **Max. Speed:** 75 m.p.h.

TRBT   19.60 x 2.84 m   37 tons   –/26

BR   *SR*
S 12529 S 12529 TRBT (ex-unit 3084)   Nene Valley Railway   Eastleigh 1938

4 DD
DOUBLE-DECK SUBURBAN UNITS
Built: 1949. Normal formation DMBT–TT–TT–DMBT.
System: 630 V DC third rail.
Traction Motors: Two EE of 185 kW (250 h.p.).   **Max. Speed:** 75 m.p.h.

DMBT   19.24 x 2.85 m   39 tons   –/121

S 13003 S DMBT (ex-unit 4002 –4902)   Hope Farm, Sellindge   Lancing 1949
S 13004 S DMBT (ex-unit 4002 –4902)   Northamptonshire Ironstone Rly   Lancing 1949

4 EPB (later Class 415)
SUBURBAN UNITS
Built: 1951–57. Normal formation DMBTO–TT–TTO–DMBTO.
System: 630 V DC third rail.
Traction Motors: Two EE507 of 185 kW (250 h.p.).   **Max. Speed:** 75 m.p.h.

| DMBTO | 19.05 x 2.82 m | 42 tons | –/82 |
| TTO | 18.90 x 2.82 m | 26 tons | –/102 |

S 14351 S DMBTO (ex-unit 5176)   Northamptonshire Ironstone Railway Eastleigh 1955
S 15354 S TTO   (ex-unit 5176)   Coventry Railway Centre   Eastleigh 1955
S 15396 S TTO   (ex-unit 5176)   Northamptonshire Ironstone Railway Eastleigh 1956
S 14352 S DMBTO (ex-unit 5176)   Northamptonshire Ironstone Railway Eastleigh 1955

Note: 15396 was originally in unit 5208.

2 EPB (later Class 416)
SUBURBAN UNITS
Built: 1959. Normal formation DMBSO–DTSO.
System: 630 V DC third rail.
Traction Motors: Two EE of 185 kW (250 h.p.).   **Max. Speed:** 75 m.p.h.

| DMBSO | 19.05 x 2.82 m | 40 tons | –/82 |
| DTSO | 18.90 x 2.82 m | 30 tons | –/92 |

S 14573 S DMBSO (ex-unit 5667–6307)   Coventry Railway Centre   Eastleigh 1959
S 16117 S DTSO   (ex-unit 5667–6307)   Coventry Railway Centre   Eastleigh 1959

6.2. PULLMAN CAR COMPANY EMU STOCK
GENERAL
Pullman cars owned by the Pullman Car Company operated as parts of EMU formations on the Southern Railway (later BR Southern Region). In addition the three Brighton Belle EMU sets were composed entirely of Pullman vehicles.

All vehicles are used as hauled stock except * – static exhibits.

6 Pul
Built: 1932. Six car sets incorporating one Pullman kitchen composite.
Normal formation: DMBTO–TTK–TCK–TPCK–TCK–DMBTO.

TPCK	20.40 x 2.77 m	43 tons	12/16

RUTH	S 264 S	TPCK	(ex-unit 2017–3042)	Venice–Simplon Orient Express	MC 1932
BERTHA	S 278 S	TPCK	(ex-unit 2012–3001)	West Coast Railway Co., Carnforth	MC 1932

5 Bel   BRIGHTON BELLE UNITS
Built: 1932. Five car all Pullman sets.
System: 630 V DC Third rail.
Formation: DMPBT–TPT–TPKF–TPKF–DMPBT.
Traction Motors: Four BTH of 167 kW (225 hp).

TPKF	20.40 x 2.77 m	42 tons	20/–
TPT	20.40 x 2.77 m	41 tons	–/56
DMPBT	20.62 x 2.77 m	62 tons	–/48

HAZEL *	S279S	TPKF	(ex-unit 2051–3051)	Black Bull, Moulton, N. Yorks	MC 1932
AUDREY	S280S	TPKF	(ex-unit 2052–3052)	Venice-Simplon Orient Express	MC 1932
GWEN	S281S	TPKF	(ex-unit 2053–3053)	Venice–Simplon Orient Express	MC 1932
DORIS	S282S	TPKF	(ex-unit 2051–3051)	Bluebell Railway	MC 1932
MONA	S283S	TPKF	(ex-unit 2053–3053)	Venice-Simplon Orient Express	MC 1932
VERA	S284S	TPKF	(ex-unit 2052–3052)	Venice-Simplon Orient Express	MC 1932
CAR No. 85	S285S	TPT	(ex-unit 2053–3053)	Venice-Simplon Orient Express	MC 1932
CAR No. 86	S286S	TPT	(ex-unit 2051–3051)	Venice-Simplon Orient Express	MC 1932
CAR No. 87*	S287S	TPT	(ex-unit 2052–3052)	Keith & Dufftown Railway	MC 1932
CAR No. 88	S288S	DMPBT	(ex-unit 2051–3051)	Venice-Simplon Orient Express	MC 1932
CAR No. 89*	S289S	DMPBT	(ex-unit 2051–3051)	Little Mill Inn, Rowarth, Derbys.	MC 1932
CAR No. 91*	S291S	DMPBT	(ex-unit 2052–3052)	Keith & Dufftown Railway	MC 1932
CAR No. 92	S292S	DMPBT	(ex-unit 2053–3053)	Venice-Simplon Orient Express	MC 1932
CAR No. 93	S293S	DMPBT	(ex-unit 2053–3053)	Venice-Simplon Orient Express	MC 1932

Note: CAR No. 89 is now named DERBYSHIRE BELLE.

6.3. LMS & CONSTITUENTS EMU STOCK

LNWR — EUSTON–WATFORD STOCK
Built: 1915. Oerlikon design. Normal formation: DMBTO–TTO–DTTO.
System: 630 V DC third rail. Used on Euston–Watford line.
Traction Motors: Four Oerlikon 179 kW (240 h.p.).
Max. Speed:

DMBTO   17.60 x 2.73 m   54.75 tonnes.   –/48

BR	LMS			
M 28249 M	28249	DMBTO	National Railway Museum	MC 1915

CLASS 502 — LIVERPOOL–SOUTHPORT STOCK
Built: 1939. Normal formation: DMBTO–TTO–DTTO (originally DTCO).
System: 630 V DC third rail.
Traction Motors: Four EE 175 kW.   **Max. Speed:** 65 m.p.h.

DMBTO	21.18 x 2.90 m	42.5 tonnes.	–/88
DTTO	21.18 x 2.90 m	25.5 tonnes.	–/79 (built as DTCO 53/25)

BR	LMS			
M 28361 M	28361	DMBTO	MoD BAD Kineton (N)	Derby 1939
M 29896 M	29896	DTTO	MoD BAD Kineton (N)	Derby 1939

CLASS 503 — MERSEY WIRRAL STOCK
Built: 1938. Normal formation: DMBTO–TTO (originally TCO)–DTTO.
System: 630 V DC third rail.
Traction Motors: 4 BTH 100 kW.   **Max. Speed:** 65 m.p.h.

DMBTO	18.48 x 2.77 m	36.5 tonnes.	–/56
TTO	17.77 x 2.77 m	20.5 tonnes.	–/58 (built as TCO 40/19)
DTTO	18.85 x 2.77 m	21.5 tonnes.	–/66

BR	LMS			
M 28690 M	28690	DMBTO	Coventry Railway Centre	Derby 1938
M 29720 M	29720	TTO	Coventry Railway Centre	Derby 1938
M 29289 M	29289	DTTO	Coventry Railway Centre	Derby 1938

MSJ&A STOCK
Built: 1931. Normal formation: DMBT–TC–DTT.
System: 1500 V DC overhead. Used on Manchester South Junction and Altrincham line until it was converted to 25 kV AC. This line is now part of the Manchester Metrolink system.
Traction Motors: **Max. Speed:** 65 m.p.h.

TC   17.60 x 2.85 m   31 tonnes.   24/72

BR	LMS	MSJ&A			
M 29666 M	29666	117	TC	Midland Railway-Butterley	MC 1931
M 29670 M	29670	121	TC	Midland Railway-Butterley	MC 1931

6.4. LNER & CONSTITUENTS EMU STOCK

NORTH EASTERN RAILWAY — DMLV
Built: 1904. Driving motor luggage van for North Tyneside line. After withdrawal from capital stock, this vehicle was used as a rail de-icing car.
System: 675 V DC third rail.
Traction Motors: **Max. Speed:**

DMLV 17.40 x 2.77 m 46.5 tonnes

BR	LNER	NER			
DE 900730	23267	3267	DMLV	Stephenson Railway Museum (N)	MC 1904

GRIMSBY & IMMINGHAM LIGHT RAILWAY — A1-1A
Built: 1915 by GCR Dukinfield.
Type: Single deck tram.
Seats: 64 + 8 tip-up.
Bogies: Brush.
Motors: 2 x 25 hp Dick Kerr DK9 of 18 kW.

14 Crich Tramway Village GCR Dukinfield 1914

CLASS 306 LIVERPOOL STREET–SHENFIELD STOCK
Built: 1949. Normal formation: DMSO–TBSO–DTSO.
System: 25 kV AC overhead (originally 1500 V DC overhead).
Traction Motors: Four Crompton Parkinson of 155 kW.
Max. Speed: 65 m.p.h.

DMSO	18.41 x 2.90 m	51.7 tonnes.	–/62
TBSO	16.78 x 2.90 m	26.4 tonnes.	–/46
DTSO	16.87 x 2.90 m	27.9 tonnes.	–/60

E 65217 E	DMSO	MoD BAD Kineton (N)		MC 1949
E 65417 E	TBSO	MoD BAD Kineton (N)		MC 1949
E 65617 E	DTSO	MoD BAD Kineton (N)		BRCW 1949

6.5. BR EMU STOCK

BR EMU power cars were numbered in the 6xxxx series starting with 61000 whilst trailer cars were numbered in the 7xxxx series. This does not apply to the APT-P or the battery EMU.

CLASS 302 BR

Built: 1958–60 for Fenchurch Street–Shoeburyness Services.
System: 25 kV AC overhead.
Original Formation: BDTSOL–MBS–TCsoL–DTS.
Formation as Rebuilt: BDTCOL–MBSO–TSOL–DTSO.
Traction Motors: Four English Electric EE 536A of 143.5 kW.
Max. Speed: 75 m.p.h.

DTSO	20.36 x 2.83 m	33.4 tonnes	–/88

| 75033 | DTSO | (ex-unit 201) | Mangapps Farm, Burnham-on-Crouch | York 1958 |
| 75250 | DTSO | (ex-unit 277) | Mangapps Farm, Burnham-on-Crouch | York 1959 |

CLASS 303 PRESSED STEEL

Built: 1959–61 for Glasgow area services.
System: 25 kV AC overhead.
Formation: DTSO–MBSO–BDTSO.
Traction Motors: Four MV of 155 kW.
Max. Speed: 75 m.p.h.

DTSO	20.18 x 2.83 m	34.4 tonnes	–/83
MBSO	20.18 x 2.83 m	56.4 tonnes	–/70
BDTSO	20.18 x 2.83 m	38.4 tonnes	–/83

61503	MBSO	(ex-unit 023)	Bo'ness & Kinneil Railway	PS 1960
75597	DTSO	(ex-unit 023)	Bo'ness & Kinneil Railway	PS 1960
75632	BDTSO	(ex-unit 032)	Bo'ness & Kinneil Railway	PS 1960

CLASS 307 BR

Built: 1954–56 for Liverpool Street–Southend Victoria Services.
System: 1500 V DC overhead. Converted 1960–61 to 25 kV AC overhead.
Original Formation: BDTBS–MS–TCsoL–DTSOL.
Formation as Rebuilt: BDTBSO–MSO–TSOL–DTCOL.
Traction Motors: Four GEC WT344 of 130 kW.
Max. Speed: 75 m.p.h.

BDTBSO	20.18 x 2.83 m	43 tonnes	–/66

| 75023 | BDTBSO (ex-unit 123) | The Pump House Steam & Transport Museum | Eastleigh 1956 |

CLASS 308 BR

Built: 1961 for Liverpool Street–Clacton stopping Services.
System: 25 kV AC overhead.
Original Formation: BDTCOL–MBS–TCsoL–DTS.
Formation as Rebuilt: BDTCOL–MBSO–TSOL–DTSO.
Traction Motors: Four English Electric EE 536A of 143.5 kW.
Max. Speed: 75 m.p.h.

BDTCOL	20.18 x 2.82 m	36.3 tonnes	24/52

| 75881 | BDTCOL (ex-unit 136) | The Pump House Steam & Transport Museum | York 1961 |

CLASS 311   CRAVENS

Built: 1967 for Glasgow "South Side electrification" extension to Gourock and Wemyss Bay.
System: 25 kV AC overhead.
Formation: DTSO(A)–MBSO–DTSO(B).
Traction Motors: Four AEI of 165 kW.
Max. Speed: 75 m.p.h.

MBSO	20.18 x 2.83 m	56.4 tonnes	–/70
DTSO(A)	20.18 x 2.83 m	34.4 tonnes	–/83

62174–977845	MBSO	(ex-unit 103)	Summerlee Heritage Park, Coatbridge	Cravens 1967
76414–977844	DTSO(A)	(ex-unit 103)	Summerlee Heritage Park, Coatbridge	Cravens 1967

CLASS 370
PROTOTYPE ADVANCED PASSENGER TRAIN (APT-P)

Built: 1978–80. Designed to run as pairs of six-car articulated units with two power cars in the middle, these electric trains featured active hydraulic tilt and proved to be a maintenance nightmare. The power cars were reasonably successful, and are partly the basis of the Class 91 electric locomotive.
System: 25 kV AC overhead.
Normal Formation of Trailer Rake: DTSOL–TSOL–TSRB–TUOL–TFOL–TBFOL.
Formation of preserved set: DTSOL–TBFOL–M–TRSB–TBFOL–DTSOL.
Traction Motors: Four ASEA LJMA 410F body mounted.
Wheel Dia: 853 mm.
Max. Speed: 125 m.p.h.

DTSOL	21.44 x 2.72 m	33.7 tonnes	–/52
TBFOL	21.20 x 2.72 m	31.9 tonnes	25/–
TRSBL	21.20 x 2.72 m	26.75 tonnes	–/28
M	20.40 x 2.72 m	67.5 tonnes	

48103	DTSOL	The Railway Age, Crewe (N)	Derby 1978
48106	DTSOL	The Railway Age, Crewe (N)	Derby 1979
48602	TBFOL	The Railway Age, Crewe (N)	Derby 1978
48603	TBFOL	The Railway Age, Crewe (N)	Derby 1978
48404	TSRBL	The Railway Age, Crewe (N)	Derby 1979
49002	M	The Railway Age, Crewe (N)	Derby 1979
49006	M	Locomotion: The NRM at Shildon	Derby 1980

CLASSES 410 & 411 (4 Bep & 4 Cep)   BR

Built: 1956–63 for Kent Coast electrification. Rebuilt 1979–1984. Class 410 (4 Bep) was later re-classified Class 412.
System: 750 V DC third rail.
Original Formation: DMBSO–TCK–TSK (4 Cep) TRB (4 Bep)–DMBSO.
Formation as Rebuilt: DMSO–TBCK–TSOL (4 Cep) TRSB (4 Bep)–DMSO.
Traction Motors: Two English Electric EE507 of 185 kW.
Max. Speed: 90 m.p.h.

DMSO	20.34 x 2.82 m	49 tonnes	–/56
TSOL	20.18 x 2.82 m	36 tonnes	–/64
TBCK	20.18 x 2.82 m	34 tonnes	24/8
TRSB	20.18 x 2.82 m	35.5 tonnes	–/24

61229	DMSO	(ex-units 7105–1537)	East Kent Light Railway	Eastleigh 1958
61230	DMSO	(ex-units 7105–1537)	East Kent Light Railway	Eastleigh 1958
61736	DMSO	(ex-units 7175–2304)	Dartmoor Railway, Okehampton	Eastleigh 1960
61737	DMSO	(ex-units 7175–2304)	Dartmoor Railway, Okehampton	Eastleigh 1960
61742	DMSO	(ex-units 7178–1589)	Dartmoor Railway, Okehampton	Eastleigh 1960
61743	DMSO	(ex-units 7178–1589)	Dartmoor Railway, Okehampton	Eastleigh 1960
61798	DMSO	(ex-units 7016–2305)	Eden Valley Railway, Warcop	Eastleigh 1961
61799	DMSO	(ex-units 7016–2305)	Eden Valley Railway, Warcop	Eastleigh 1961
61804	DMSO	(ex-units 7019–2301)	Eden Valley Railway, Warcop	Eastleigh 1961
61805	DMSO	(ex-units 7019–2301)	Eden Valley Railway, Warcop	Eastleigh 1961
69013	TRSB	(ex-units 7014–2305)	East Kent Light Railway	Eastleigh 1961

70229	TSOL	(ex-units 7105–1537)	Eden Valley Railway, Warcop	Eastleigh	1958
70235	TBCK	(ex-units 7105–1537)	East Kent Light Railway	Eastleigh	1958
70262	TSOL	(ex-units 7113–1524)	St. Leonards Railway Engineering	Eastleigh	1958
70273	TSOL	(ex-units 7124–1530)	Dean Forest Railway	Eastleigh	1958
70284	TSOL	(ex-units 7135–1520)	Northamptonshire Ironstone Railway	Eastleigh	1959
70292	TSOL	(ex-units 7143–1554)	West Bay Visitor Centre, near Bridport	Eastleigh	1959
70294	TSOL	(ex-units 7145–1552)	HM Prison Lindholme, near Doncaster	Eastleigh	1959
70296	TSOL	(ex-units 7147–1559)	Northamptonshire Ironstone Railway	Eastleigh	1959
70300	TSOL	(ex-units 7151–1540)	Fighting Cocks PH, Middleton St. George	Eastleigh	1959
70345	TBCK	(ex-units 7153–1500)	Hydraulic House, Sutton Bridge	Eastleigh	1959
70346	TBCK	(ex-units 7003–1532)	Angel Entertainment, Long Marston Aerodrome	Eastleigh	1959
70354	TBCK	(ex-units 7011–2305)	Eden Valley Railway, Warcop	Eastleigh	1959
70508	TSOL	(ex-units 7159–1595)	Dartmoor Railway, Okehampton	Eastleigh	1960
70510	TSOL	(ex-units 7161–1597)	Northamptonshire Ironstone Railway	Eastleigh	1960
70512	TSOL	(ex-units 7163–1605)	HM Prison Lindholme, near Doncaster	Eastleigh	1960
70527	TSOL	(ex-units 7178–1589)	Great Central Railway	Eastleigh	1960
70531	TSOL	(ex-units 7152–1610)	West Bay Visitor Centre, near Bridport	Eastleigh	1961
70539	TSOL	(ex-units 7190–1568)	Eden Valley Railway, Warcop	Eastleigh	1961
70547	TSOL	(ex-units 7198–1569)	Garden Art, Bath Road, Hungerford	Eastleigh	1961
70549	TSOL	(ex-units 7200–1567)	East Lancashire Railway	Eastleigh	1961
70573	TBCK	(ex-units 7175–2304)	Dartmoor Railway, Okehampton	Eastleigh	1960
70576	TBCK	(ex-units 7178–1589)	Snibston Discovery Park	Eastleigh	1960
70607	TBCK	(ex-units 7019–2301)	Eden Valley Railway, Warcop	Eastleigh	1961

Note: 69013 also carried the number 69345 after rebuild as a TRSB.

CLASS 414 (2 Hap)   BR

Built: 1959 for the South-Eastern Division of the former BR Southern Region.
System: 750 V DC third rail.
Formation: DMBSO–DTCsoL.
Traction Motors: Two English Electric EE507 of 185 kW.
Max. Speed: 90 m.p.h.

DMBSO	20.44 x 2.82 m	42 tonnes	–/84
DTCsoL	20.44 x 2.82 m	32.5 tonnes	19/60

61275	DMBSO	(ex-units 6077–4308)	National Railway Museum	Eastleigh	1959
61287	DMBSO	(ex-units 6089–4311)	Coventry Railway Centre	Eastleigh	1959
75395	DTCsoL	(ex-units 6077–4308)	National Railway Museum	Eastleigh	1959
75407	DTCsoL	(ex-units 6089–4311)	Coventry Railway Centre	Eastleigh	1959

CLASS 416 (2 EPB)   BR

Built: 1955–56. 65373 & 77558 were built for the South Eastern Division of BR Southern Region. 65321 & 77112 were originally used between Newcastle and South Shields but were transferred to join the rest of the class on the Southern Region when the South Tyneside line was de-electrified in 1963.
System: 750 V DC third rail.
Formation: DMBSO–DTSso.
Traction Motors: Two English Electric EE507 of 185 kW.
Max. Speed: 75 m.p.h.

DMBSO	20.44 x 2.82 m	42 tonnes	–/82
DTSso	20.44 x 2.82 m	30.5 tonnes	–/102

65321–977505	DMBSO	(ex-unit 5791–6291)	Coventry Railway Centre	Eastleigh	1955
65373	DMBSO	(ex-units 5759–6259)	East Kent Light Railway	Eastleigh	1956
77112–977508	DTSso	(ex-units 5793–6293)	Coventry Railway Centre	Eastleigh	1955
77558	DTSso	(ex-unit 5759–6259)	East Kent Light Railway	Eastleigh	1956

CLASS 419 (MLV) BR

Built: 1959–61. Motor luggage vans for Kent Coast electrification. Fitted with traction batteries to allow operation on non-electrified lines.
System: 750 V DC third rail.
Traction Motors: Two English Electric EE507 of 185 kW.
Max. Speed: 90 m.p.h.
Also fitted with vacuum brakes for hauling parcels trains.

DMLV   20.45 x 2.82 m   45.5 tonnes

68001	(ex-units 9001, 931 091)	East Kent Light Railway	Eastleigh 1959
68002	(ex-units 9002, 931 092)	East Kent Light Railway	Eastleigh 1959
68003	(ex-units 9003, 931 093)	Eden Valley Railway, Warcop	Eastleigh 1961
68004	(ex-units 9004, 931 094)	Mid Norfolk Railway	Eastleigh 1961
68005	(ex-units 9005, 931 095)	Eden Valley Railway, Warcop	Eastleigh 1961
68008	(ex-units 9008, 931 098)	East Kent Light Railway	Eastleigh 1961
68009	(ex-units 9009, 931 099)	Colne Valley Railway	Eastleigh 1961
68010	(ex-units 9010, 931 090)	Wensleydale Railway	Eastleigh 1961

CLASS 420 & 421 (4 Big & 4 Cig) BR

Built: 1963–69 for the Central Division of BR Southern Region. Class 420 (4 Big) was later reclassified Class 422.
System: 750 V DC third rail.
Formation: DTCsoL–MBSO–TSO (4 Cig), TSRB (4 Big)–DTCsoL.
Traction Motors: Two English Electric EE507 of 185 kW.
Max. Speed: 90 m.p.h.

MBSO	20.18 x 2.82 m	49 tonnes	–/56
TSO	20.18 x 2.82 m	31.5 tonnes	–/72
DTCso	20.23 x 2.82 m	35 tonnes	24/28
TSRB	20.18 x 2.82 m	35 tonnes	–/40

62043 MBSO	(ex-units 7327–1127–1753)	Ellough Airfield, near Beccles	York 1965
62287 MBSO	(ex-units 7337–1237–1816–1303)	Lincolnshire Wolds Railway	York 1970
62364 MBSO	(ex-units 7376–1276–2251–1374)	Dean Forest Railway	York 1971
62378 MBSO	(ex-units 7390–1290–2255–1392)	Dean Forest Railway	York 1971
62384 MBSO	(ex-units 7396–1296–2258–1353)	Great Central Railway	York 1971
62385 MBSO	(ex-units 7397–1297–2256–1399)	Dartmoor Railway, Okehampton	York 1971
69302 TSRB	(ex-units 7032–1276–2251)	Abbey View Disabled Day Centre	York 1963
69304 TSRB	(ex-units 7034–1299–2260)	Northamptonshire Ironstone Railway	York 1963
69306 TSRB	(ex-units 7036–1282–2254)	Spa Valley Railway	York 1963
69310 TSRB	(ex-units 7040–1290–2255)	Dartmoor Railway, Okehampton	York 1963
69316 TSRB	(ex-units 7046–1296–2258)	Waverley Route Heritage Association	York 1963
69318 TSRB	(ex-units 7048–1298–2259)	Colne Valley Railway	York 1963
69332 TSRB	(ex-units 7051–2203)	Dartmoor Railway, Okehampton	York 1969
69333 TSRB	(ex-units 7055–1802–2260)	Lavender Line, Isfield	York 1969
69335 TSRB	(ex-units 7057–2209)	Wensleydale Railway	York 1969
69337 TSRB	(ex-units 7058–2210)	St. Leonards Railway Engineering	York 1969
69338 TSRB	(ex-units 7054–2206)	Station Restaurant, Gulf Corporation, Bahrain	York 1969
69339 TSRB	(ex-units 7053–2205)	Great Central Railway	York 1969
70721 TSO	(ex-units 7327–1127–1753)	Ellough Airfield, near Beccles	York 1965
71041 TSO	(ex-units 7373–1273–1819–1306)	Hever Station, Kent	York 1971
76048 DTCso	(ex-units 7327–1127–1753)	Ellough Airfield, near Beccles	York 1965
76102 DTCso	(ex-units 7327–1127–1753)	Ellough Airfield, near Beccles	York 1965
76726 DTCso	(ex-units 7376–1276–2251–1374)	Dean Forest Railway	York 1971
76740 DTCso	(ex-units 7390–1290–2255–1392)	Dean Forest Railway	York 1971
76746 DTCso	(ex-units 7396–1296–2258–1393)	Great Central Railway	York 1971
76747 DTCso	(ex-units 7397–1297–2256–1399)	Dartmoor Railway, Okehampton	York 1971
76797 DTCso	(ex-units 7376–1276–2251–1374)	Dean Forest Railway	York 1971
76811 DTCso	(ex-units 7390–1290–2255–1392)	Dean Forest Railway	York 1971
76817 DTCso	(ex-units 7396–1296–2258–1353)	Great Central Railway	York 1971
76818 DTCso	(ex-units 7397–1297–2256–1399)	Dartmoor Railway, Okehampton	York 1971

CLASS 423 (4 Vep) BR
Built: 1967–74 for BR Southern Region.
System: 750 V DC third rail.
Formation: DTCsoL–MBSO–TSO–DTCsoL.
Traction Motors: Four English Electric EE507 of 185 kW.
Max. Speed: 90 m.p.h.

DTCso	20.18 x 2.82 m	35.0 tonnes	18/46
TSO	20.18 x 2.82 m	31.5 tonnes	–/98
MBSO	20.18 x 2.82 m	48.0 tonnes	–/76

62351	MBSO	(ex-units 7850–3150–3504–3822)	Churnet Valley Railway	York 1973
71032	TSO	(ex-units 7851–3151–3504–3822)	Churnet Valley Railway	York 1973
76350	DTCso	(ex-units 7729–3029–3436)	Bideford Station Museum	York 1967
76529	DTCso	(ex-units 7860–3160–3438–3414–3822)	Churnet Valley Railway	York 1969
76712	DTCso	(ex-units 7851–3151–3504–3822)	Churnet Valley Railway	York 1973
76875	DTCso	(ex-units 7861–3161–3595)	National Railway Museum	York 1973
76887	DTCso	(ex-units 7867–3167–3568)	Woking Miniature Railway	York 1973

CLASS 489 (GLV) BR
Rebuilt: 1979–84 from Class 414 (2 Hap) as Motor Luggage Vans for the Victoria–Gatwick "Gatwick Express" service. Seats removed.
System: 750 V DC third rail.
Traction Motors: Two English Electric EE507 of 185 kW.
Max. Speed: 90 m.p.h.

DMLV	20.04 x 2.82 m	45 tonnes	

61267–68507	GLV	(ex-units 6069–9108)	Nottingham Heritage Centre	Eastleigh 1959
61269–68500	GLV	(ex-units 6071–9101)	Ecclesbourne Valley Railway	Eastleigh 1959
61277–68503	GLV	(ex-units 6079–9104)	Spa Valley Railway	Eastleigh 1959
61280–68509	GLV	(ex-units 6082–9110)	Barry Island Railway	Eastleigh 1959
61292–68506	GLV	(ex-units 6094–9117)	Ecclesbourne Valley Railway	Eastleigh 1959

CLASS 491 (4 TC) BR/METRO-CAMMELL
Built: 1967. Unpowered units designed to work push-pull with Class 430 (4 Rep) tractor units and Class 33/1, 73 and 74 locomotives. Converted from locomotive hauled coaching stock built 1952–57 (original number in brackets).
System: 750 V DC third rail.
Formation: DTSO–TFK–TBSK–DTSO.
Max. Speed: 90 m.p.h.

DTSO	20.18 x 2.82 m	32 tonnes	–/64
TFK	20.18 x 2.82 m	33.5 tonnes	42/–
TBSK	20.18 x 2.82 m	35.5 tonnes	–/32

70812 (34987)	TBSK	(ex-unit 401)	Dartmoor Railway, Okehampton	MC 1957
70823 (34970)	TBSK	(ex-unit 412)	London Underground, West Ruislip Depot	MC 1957
70824 (34984)	TBSK	(ex-unit 413)	Westfield Storage, Dinton	MC 1957
70826 (34980)	TBSK	(ex-unit 415)	Dartmoor Railway, Okehampton	MC 1957
70855 (13018)	TFK	(ex-unit 412)	Westfield Storage, Dinton	Swindon 1952
70859 (13040)	TFK	(ex-unit 412)	Stravithie Station, Fife	Swindon 1952
70860 (13019)	TFK	(ex-unit 417)	Dartmoor Railway, Okehampton	Swindon 1952
71163 (13097)	TFK	(ex-unit 430)	London Underground, West Ruislip Depot	Swindon 1954
76275 (3929)	DTSO	(ex-unit 404)	St. Leonards Railway Engineering	Eastleigh 1955
76277 (4005)	DTSO	(ex-unit 405)	Dartmoor Railway, Okehampton	Swindon 1957
76297 (3938)	DTSO	(ex-unit 415)	London Underground, West Ruislip Depot	Eastleigh 1955
76298 (4004)	DTSO	(ex-unit 415)	Westfield Storage, Dinton	Eastleigh 1957
76301 (4375)	DTSO	(ex-unit 417)	Dartmoor Railway, Okehampton	Swindon 1957
76302 (4382)	DTSO	(ex-unit 417)	Dartmoor Railway, Okehampton	Swindon 1957
76322 (3936)	DTSO	(ex-unit 427)	Westfield Storage, Dinton	Eastleigh 1955
76324 (4009)	DTSO	(ex-unit 428)	London Underground, West Ruislip Depot	Eastleigh 1957

CLASS 501   BR

Built: 1957 for Euston–Watford and North London lines.
Normal Formation: DMBSO–TSO–DTBSO.
System: 630 V DC third rail.
Traction Motors: Two English Electric GEC of 135 kW.
Max. Speed: 60 m.p.h.

| DMBSO | 18.47 x 2.82 m | 47.8 tonnes | –/74 |
| DTBSO | 18.47 x 2.82 m | 30.5 tonnes | –/74 |

BR	AD			
61183–DB 975349		DMBSO	Coventry Railway Centre	Eastleigh 1957
75186	WGP 8809	DTBSO	Coventry Railway Centre	Eastleigh 1957

CLASS 504   BR

Built: 1959 for Manchester–Bury line.
Normal Formation: DMBSO–DTSO.
System: 1200 V DC protected side-contact third rail.
Max. Speed: 65 m.p.h.
Traction Motors: Four English Electric EE of 90 kW.

| DMBSO | 20.31 x 2.82 m | 50 tonnes | –/84 |
| DTSO | 20.31 x 2.82 m | 33 tonnes | –/94 |

| 65451 | DMBSO | East Lancashire Railway | Wolverton 1959 |
| 77172 | DTSO | East Lancashire Railway | Wolverton 1959 |

TRAMS   GRIMSBY & IMMINGHAM LIGHT RAILWAY

Built: 1925.(1927*). Single-deck trams built for Gateshead & District Tramways Company.
Sold to British Railways in 1951 for use on Grimsby & Immingham Light Railway.
Motors: 2 x Dick Kerr 31A (25A*) of 18 kW (25 h.p.).
Bogies: Brill 39E.
Seats: 48 (reduced to 44 on BR*).

BR	Gateshead		
20 *	5	Crich Tramway Village	Gateshead & District Tramways Co. 1927
26	10	North of England Open Air Museum	Gateshead & District Tramways Co. 1925

BATTERY EMU   BR DERBY/COWLAIRS TWIN UNIT

Built: 1958. Normal formation. BDMBSO–BDTCOL.
Power: 216 lead-acid cells of 1070 Ah.
Traction Motors: Two 100 kW Siemens nose-suspended motors.
Max. Speed: 70 m.p.h.
Braking: Vacuum

| BDMBSO | 18.49 x 2.79 m | 37.5 tonnes | –/52 |
| BDTCOL | 18.49 x 2.79 m | 32.5 tonnes | 12/53 |

| 79998–DB 975003 | BDMBS | Royal Deeside Railway | Derby/Cowlairs 1958 |
| 79999–DB 975004 | BDTCL | Royal Deeside Railway | Derby/Cowlairs 1958 |

7. LIST OF LOCATIONS

The following is a list of locations in Great Britain where locomotives and multiple units included in this book can be found, together with Ordnance Survey grid references where these are known. At certain locations locomotives may be dispersed at several sites. In such cases the principle site where locomotives can normally be found is the one given. Enquiries at this site will normally reveal the whereabouts of other locomotives, but this is not guaranteed.

7.1. PRESERVATION SITES & OPERATING RAILWAYS

§ denotes not normally open to the public.

	OS GRID REF.
A1 Loco Trust, Former Carriage Works, Hopetown, Darlington, Co. Durham§	NZ 288157
Abbey View Disabled Day Centre, Neath Abbey, Neath, Port Talbot.	SS 734973
Allely's, The Slough, Studley, Warwickshire.§	SP 057637
Amman Valley Railway, TDW Distribution, Fabian Way, Port Tennant, Swansea.§	SS 682931
Angel Entertainment, Long Marston Aerodrome, Long Marston, Warwickshire.	SP 170486
Appleby Heritage Centre, Appleby, Cumbria.	NY 688205
Appleby-Frodingham RPS, Corus, Appleby-Frodingham Works, Scunthorpe, Lincs.	SE 913109
Avon Valley Railway, Bitton Station, Bitton, Gloucestershire.	ST 670705
Barclay Brown Yard, High Street, Methil, Fife.§	NT 372994
Barrow Hill Roundhouse, Campbell Drive, Barrow Hill, Staveley, Chesterfield, Derbyshire.	SK 414755
Barry Island Railway, Barry Island, Vale of Glamorgan.	ST 118667
Battlefield Railway, Shackerstone Station, Shackerstone, Leicestershire.	SK 379066
Bideford Station Museum, Bideford, Devon.	SS 458264
Birmingham Museum of Science & Industry, Discovery Centre, Millennium Point, Birmingham.	SP 079873
Bluebell Railway, Sheffield Park, Uckfield, East Sussex.	TQ 403238
Bodmin Steam Railway, Bodmin General Station, Bodmin, Cornwall.	SX 073664
Bo'ness & Kinneil Railway, Bo'ness Station, Union Street, Bo'ness, Falkirk.	NT 003817
Bressingham Steam Museum, Bressingham Hall, Diss, Norfolk.	TM 080806
Brighton Railway Museum, Preston Park, Brighton, West Sussex.§	TQ 302061
Bryn Engineering, Blackrod Industrial Estate, Scot Lane, Blackrod, near Bolton.§	SD 623089
Bryn Engineering, Locomotive Restoration Centre, Pemberton, Wigan, Greater Manchester.§	SD 550036
Buckinghamshire Railway Centre, Quainton Road Station, Aylesbury, Bucks.	SP 736189
Caledonian Railway, Brechin Station, Brechin, Montrose, Angus.	NO 603603
Cambrian Railway Trust, Llynclys, near Oswestry, Shropshire.	SJ 284239
Canterbury Heritage Centre, Stour Street, Canterbury, Kent.	TQ 146577
Cefn Coed Colliery Museum, Old Blaenant Colliery, Cryant, Neath Port Talbot.	SN 786034
Chasewater Light Railway, Chasewater Pleasure Park, Brownhills, Staffordshire.	SK 034070
Chinnor & Princes Risborough Railway, Chinnor Cement Works, Chinnor, Oxon.	SP 756002
Cholsey & Wallingford Railway, St. John's Road, Wallingford, Oxfordshire.	SU 600891
Churnet Valley Railway, Cheddleton Station, Cheddleton, Leek, Staffordshire.	SJ 983519
Colne Valley Railway, Castle Hedingham Station, Halstead, Essex.	TL 774362
Cotswold Rail, Gloucester Depot, Horton Road, Gloucester, Gloucestershire§	SO 842183
Coventry Railway Centre, Rowley Road, Baginton, Coventry, Warwickshire.	SP 354751
Crich Tramway Village, Crich, nr Matlock, Derbyshire.	SK 345549
Darlington North Road Goods Shed, Station Road, Hopetown, Darlington, Co. Durham.	NZ 290157
Darlington North Road Museum, Station Road, Hopetown, Darlington, Co. Durham.	NZ 289157
Dartmoor Railway, Meldon Quarry, Okehampton, Devon.	SX 568927
Dean Forest Railway, Norchard, Lydney, Gloucestershire.	SO 629044
Denbigh & Mold Junction Railway, Sodom, near Bodfari, Denbighshire, Wales.	SJ 103711
Derwent Valley Light Railway, Yorkshire Museum of Farming, Murton, York, North Yorks.	SE 650524
Designer Outlet Village, Kemble Drive, Swindon, Wiltshire.	SU 142849
Didcot Railway Centre (Great Western Society), Didcot, Oxfordshire.	SU 524906

Ecclesbourne Valley Railway, Wirksworth Station, Wirksworth, Derbyshire.	SK 289542
East Anglian Railway Museum, Chappel & Wakes Colne Station, Essex.	TL 898289
East Kent Light Railway, Shepherdswell, Kent.	TR 258483
East Lancashire Railway, Bolton Street Station, Bury, Greater Manchester.	SD 803109
East Somerset Railway, West Cranmore Station, Shepton Mallet, Somerset.	ST 664429
Eden Valley Railway, Warcop Station, Warcop, Cumbria.	NY 758152
Ellough Airfield, near Beccles, Suffolk.§	
Embsay & Bolton Abbey Railway, Embsay Station, Embsay, Skipton, North Yorks.	SE 007533
Enfield Timber Company, The Ridge, Hertford Road, Enfield, Middlesex.§	TQ 352965
Epping & Ongar Railway, Ongar Station, Station Road, Chipping Ongar, Essex.	TL 552035
Essex Traction Group, Boxted, near Colchester, Essex.§	
Fawley Hill Railway, Fawley Green, near Henley-on-Thames, Buckinghamshire.§	SU 755861
Fire Service College, Moreton-in-Marsh, Gloucestershire.§	SP 216329
Foxfield Railway, Blythe Bridge, Stoke-on-Trent, Staffordshire.	SJ 976446
Garden Art, 1 Bath Road, Hungerford, Berkshire.	SU 343689
GCE & SCS, Tradecroft Industrial Estate, Easton, Isle of Portland, Dorset.	SY 685723
Glasgow Museum of Transport, Kelvin Hall, Burnhouse Road, Coplawhill, Glasgow.	NS 565663
Glasgow Museums Resource Centre, South Nitshill, Glasgow.	
Gloucestershire-Warwickshire Railway, Toddington Station, Gloucestershire.	SP 049321
Great Central Railway, Loughborough Central Station, Loughborough, Leicestershire.	SK 543194
Greater Manchester Museum of Science & Industry, Liverpool Road, Manchester.	SJ 831978
Gwili Railway, Bronwydd Arms Station, Carmarthen, Carmarthenshire.	SN 417236
GWR Preservation Group, Field Sidings, Southall, Greater London.	TQ 133798
Helston Railway, Gwinear Road Station, near Cambourne, Cornwall.	SW 613384
Hever Station, Hever, Kent.§	TQ 465445
HLPG, Binbrook Trading Estate, Binbrook, Lincolnshire.§	TF 201958
HM Prison Lindholme, Moor Dike Road, Lindholme, near Doncaster.§	SE 687057
Hope Farm (Southern Locomotives), Sellindge, near Ashford, Kent.§	TR 119388
Hydraulic House, West Bank, Sutton Bridge, near Spalding, Lincolnshire.§	TF 488242
Ian Storey Engineering, Station Yard, Hepscott, Morpeth, Northumberland.§	NZ 223844
Isle of Wight Steam Railway, Haven Street Station, Isle of Wight.	SZ 556898
Keighley & Worth Valley Railway, Haworth, Keighley, West Yorkshire.	SE 034371
Keith & Dufftown Railway, Dufftown, Moray.	NJ 323414
Kent & East Sussex Railway, Tenterden Town Station, Tenterden, Kent.	TQ 882336
Kingdom of Fife Railway, Old Kirkland Yard, Burnmills Industrial Estate, Leven, Fife.§	NO 373008
Lakeside & Haverthwaite Railway, Haverthwaite, Cumbria.	SD 349843
Lavender Line, Isfield Station, Station Road, Isfield, East Sussex.	TQ 452171
Leeds Industrial Museum, Armley Mills, Canal Road, Leeds, West Yorkshire.	SE 275342
Lincolnshire Wolds Railway, Ludborough Station, Ludborough, Lincolnshire.	TF 309960
Liverpool Museum Store, Juniper Street, Bootle, Merseyside.§	SJ 343935
Llanelli & Mynydd Mawr Railway, Cynheidre, near Llanelli, Carmarthenshire.	SN 49x07x
Llangollen Railway, Llangollen Station, Llangollen, Denbighshire.	SJ 211423
LNWR Crewe Carriage Shed, Crewe, Cheshire.§	SJ 715539
Locomotion: The NRM at Shildon, Shildon, County Durham.	NZ 238256
London Transport Depot Museum, Gunnersby Lane, Acton, Greater London.	TQ 194799
London Underground, West Ruislip Depot, Ruislip, London.§	TQ 094862
Long & Somerville, Southbury Road, Potters End, Enfield, London.§	TQ 348962
Long Marston Workshops, Long Marston, Warwickshire.§	SP 152469
Mangapps Farm, Southminster Road, Burnham-on-Crouch, Essex.	TQ 944980
Marshalls Transport, Pershore Airfield, Throckmorton, near Evesham, Worcestershire.§	SO 978506
Middleton Railway, Tunstall Road, Hunslet, Leeds, West Yorkshire.	SE 305310
Mid Hants Railway, Alresford Station, New Alresford, Hampshire.	SU 588325
Midland Railway-Butterley, Butterley Station, Ripley, Derbyshire.	SK 403520
Mid-Norfolk Railway, Dereham Station, Dereham, Norfolk.	TG 003102
MoD Ashchurch, Gloucestershire.§	SO 932338
MoD BAD Kineton, Temple Herdewyke, Warwickshire.§	SO 402291
MoD Llangennech, Glanmwrwg, Llangennech, Carmarthenshire.§	SN 557017
Moor Street Station, Birmingham, West Midlands.	SP 074868
Moreton Park Railway, Moreton-on-Lugg, near Hereford, Herefordshire.§	SO 503467
National Railway Museum, Leeman Road, York, North Yorkshire.	SE 594519
NELPG, Former Carriage Works, Hopetown, Darlington, County Durham.	NZ 288157
Nene Valley Railway, Wansford Station, Peterborough, Cambridgeshire.	TL 093979
Northampton & Lamport Railway, Pitsford, Northamptonshire.	SP 736666

Northamptonshire Ironstone Railway, Hunsbury Hill, Northampton, Northamptonshire.	SP 735584
North Dorset Railway, Shillingstone Station, Shillingstone, Dorset.	ST 824117
North Norfolk Railway, Sheringham Station, Norfolk.	TG 156430
North of England Open Air Museum, Beamish Hall, Beamish, County Durham.	NZ 217547
North Woolwich Station Museum, Pier Road, North Woolwich, London.	TQ 433798
North Yorkshire Moors Railway, Grosmont Station, North Yorkshire.	NZ 828049
Nottingham Heritage Centre, Mereway, Ruddington, Nottinghamshire.	SK 575322
Old Oak Common Depot, Acton, Greater London.§	TQ 218823
Oswestry Cycle & Railway Museum, Oswestry Station Yard, Oswestry, Shropshire.	SJ 294297
Paignton & Dartmouth Railway, Queen's Park Station, Paignton, Devon.	SX 889606
Peak Railway, Darley Dale Station, Darley Dale, Matlock, Derbyshire.	SK 274626
Plym Valley Railway, Marsh Mills, Plymouth, Devon.	SX 520571
Pontypool & Blaenavon Railway, Furnoe Sidings, Big Pit, Blaenavon, Torfaen.	SO 237093
Pullman Design & Fabrication, Canton, Cardiff.§	ST 177779
The Pump House Steam & Transport Museum, The Pump House, Lowe Hall Lane, Walthamstow, London.	TQ 362882
The Railway Age, Crewe, Cheshire.	SJ 708552
Railworld (Museum of World Railways), Woodston, Peterborough, Cambs.	TL 188982
Reliance Industrial Estate, Eager Street, Newton Heath, Manchester.§	SD 873009
Ribble Steam Railway, off Chain Caulway, Riversway, Preston, Lancashire.	SD 504295
Rippingale Station, Fen Road, Rippingale, Lincolnshire.§	TF 115283
Rogart Station, Rogart, Highland.	NC 724020
Rowden Mill Station, Rowden Mill, near Bromyard, Herefordshire.§	SO 627565
Royal Deeside Railway, Milton, Crathes, Banchory, Aberdeenshire.	NO 743962
Rushden Station Museum, Rectory Road, Rushden, Northamptonshire.	SP 957672
Rutland Railway Museum, Ashwell Road, Cottesmore, Oakham, Rutland.	SK 887137
RVEL, rtc Business Park, London Road, Derby, Derbyshire.§	SK 365350
Rye Farm, Ryefield Lane, Wishaw, Sutton Coldfield, Warwickshire.	SP 180944
Science Museum, Imperial Institute Road, South Kensington, London.	TQ 268793
Scolton Manor Museum, Scolton Manor, Haverfordwest, Pembrokeshire.	SM 991222
Severn Valley Railway, Bridgnorth Station, Shropshire.	SO 715926
Seward Agricultural Machinery Suppliers, Sinderby, near Thirsk, North Yorkshire.§	SE 334813
Snibston Discovery Park, Snibston Mine, Ashby Road, Coalville, Leicestershire.	SK 420144
South Devon Railway, Buckfastleigh, Devon.	SX 747663
Southall Depot, Southall, Greater London.§	TQ 133798
Spa Valley Railway, Tunbridge Wells West Station, Tunbridge Wells, Kent.	TQ 542346
Spirit of The West American Theme Park, Retallack, St Columb Major, Cornwall.	SW 936658
St. Leonards Railway Engineering, West Marina Depot, Bridge Way, St Leonards, East Sussex.§	TQ 778086
St. Modwen Properties, Long Marston, Warwickshire.§	SP 152469
Stainmore Railway, Kirkby Stephen East Station, Kirkby Stephen, Cumbria.	NY 769075
Steam-Museum of the Great Western Railway, Old No. 20 Shop, Old Swindon Works, Kemble Drive, Swindon, Wiltshire.	SU 143849
Steam Traction, Acton Place Industrial Estate, Acton, Sudbury, Suffolk.§	TL 883455
Stephenson Railway Museum, Middle Engine Lane, West Chirton, Tyne & Wear.	NZ 323693
Stewarts Lane Depot, Battersea, London.§	TQ 257798
Strathspey Railway, Aviemore, Highland Region.	NH 898131
Stravithie Station, Stravithie, near St. Andrews, Fife.	NO 533134
Summerlee Heritage Park, West Canal Street, Coatbridge, North Lanarkshire.	NS 728655
Swanage Railway, Swanage Station, Swanage, Dorset.	SZ 028789
Swansea Vale Railway, Llansamlet, Swansea.	SS 660928
Swindon & Cricklade Railway, Blunsden Road Station, Swindon, Wiltshire.	SU 110897
Swindon Railway Workshop, Flour Mill Engine House, Bream, Forest of Dean, Glos.§	SO 604067
Talyllyn Railway, Tywyn Pendre Depot, Tywyn, Gwynedd.	SH 590008
Tanfield Railway, Marley Hill Engine Shed, Sunniside, Tyne & Wear.	NZ 207573
Telford Steam Railway, Bridge Road, Horsehay, Telford, Shropshire.	SJ 675073
Thornton Depot, Strathore Road, Thornton, Fife, Scotland.§	NT 264970
Titley Junction Station, near Kington, Herefordshire.§	SO 329581
Tiverton Museum, St. Andrew's Street, Tiverton, Devon.	SS 955124
Tyseley Locomotive Works, Warwick Road, Tyseley, Birmingham.	SP 105841
Vale of Rheidol Railway, Aberystwyth, Ceredigian.	SN 587812
Venice-Simplon Orient Express, Stewarts Lane Depot, Battersea, London.§	TQ 288766
Waverley Route Heritage Association, Whitrope, near Hawick, Scottish Borders.	NT 527005

Weardale Railway, Wolsingham, County Durham.	NZ 081370
Welshpool & Llanfair Railway, Llanfair Caereinon, Powys.	SJ 107069
Wembley Depot, Wembley, Greater London.§	TQ 193844
Wensleydale Railway, Leeming Bar Station, Leeming Bar, North Yorkshire.	SE 286900
West Bay Visitor Centre, West Bay, Bridport, Dorset.	SY 466904
West Coast Railway Company, Warton Road, Carnforth, Lancashire.§	SD 496708
West Somerset Railway, Minehead Station, Minehead, Somerset.	SS 975463
Westfield Storage, Dunton, Wiltshire.§	
Woking Miniature Railway, Barr's Lane, Knaphill, Woking, Surrey.	TQ 966595
Wrenbury Business Park, Wrenbury, near Nantwich, Cheshire.§	SJ 603472
Yeovil Railway Centre, Yeovil Junction, near Yeovil, Somerset.	ST 271141

7.2. PUBLIC HOUSES & HOTELS

Black Bull, Moulton, North Yorkshire.	NZ 237037
Fighting Cocks, Middleton St. George, Near Darlington, Co. Durham.	NZ 342143
Little Mill Inn, Rowarth, Mellor, Derbyshire.	SK 011890

8. ABBREVIATIONS USED

ABB	ASEA Brown Boveri
AC	Alternating Current
AD	Army Department of the Ministry of Defence
AD	Alexandra Docks & Railway Company
BAD	Base Armament Depot
BR	British Railways
BPGVR	Burry Port & Gwendraeth Valley Railway
BREL	British Rail Engineering Ltd. (later BREL, then ABB, now Bombardier Transportation).
BTH	British Thomson Houston
C&W	Carriage & Wagon
CR	Caledonian Railway
DC	Direct Current
DE	Diesel electric
DH	Diesel hydraulic
DM	Diesel mechanical
DMU	Diesel Multiple Unit
OW&W	Oxford, Worcester & Wolverhampton Railway
FR	Furness Railway
FS	Ferrovie dello Stato
GCR	Great Central Railway
GEC	General Electric Company (UK)
GER	Great Eastern Railway
GJR	Grand Junction Railway
GNR	Great Northern Railway
GNSR	Great North of Scotland Railway
GSWR	Glasgow & South Western Railway
GVR	Gwendraeth Valley Railway
GWR	Great Western Railway
HLPG	Humberside Locomotive Preservation Society
HR	Highland Railway
IRR	Iran Islamic Republic Railways
L&MR	Liverpool & Manchester Railway
LBSCR	London, Brighton & South Coast Railway
LC&W	Locomotive, Carriage & Wagon
LMS	London Midland & Scottish Railway
LNER	London & North Eastern Railway
LNWR	London & North Western Railway
LSWR	London & South Western Railway
L&Y	Lancashire & Yorkshire Railway
LTE	London Transport Executive
LTSR	London Tilbury & Southend Railway
MoD	Ministry of Defence

MSJ&A	Manchester South Junction & Altrincham Railway
MR	Midland Railway
NBR	North British Railway
NELPG	North Eastern Locomotive Preservation Group
NER	North Eastern Railway
NLR	North London Railway
NS	Nederlandse Spoorwegen
NSR	North Staffordshire Railway
P&M	Powlesland & Mason
PT	Pannier Tank
PTR	Port Talbot Railway
RC&W	Railway Carriage & Wagon
RLC	Royal Logistics Corps
Rly	Railway
ROD	Railway Operating Department
RPS	Railway Preservation Society
RTC	Railway Technical Centre
S&DJR	Somerset & Dorset Joint Railway
SDR	South Devon Railway
SECR	South Eastern & Chatham Railway
SJ	Swedish State Railways
SPR	Sandy & Potton Railway
SR	Southern Railway
ST	Saddle Tank
T	Side Tank
TCDD	Türkiye Cumhuryeti devlet Demiryollan
TVR	Taff Vale Railway
USATC	United States of America Transportation Corps
W&L	Welshpool & Llanfair Railway
WD	War Department
WR	British Railways Western Region
WT	Well Tank
WT	Wantage Railway
(S)	Stored

9. PRIVATE MANUFACTURER CODES

The following codes are used to denote private locomotive manufacturers. These are followed by the works number and build year, e.g. AW 1360/1937 – built by Armstrong-Whitworth and Company, works number 1360, year 1937. Unless otherwise shown, locations are in England.

AB	Andrew Barclay, Sons & Company, Caledonia Works, Kilmarnock, Scotland.
AC	AC Cars, Thames Ditton, Surrey.
AE	Avonside Engine Company, Bristol, Avon.
AEC	Associated Equipment Company, Southall, Berkshire.
AL	American Locomotive Company, USA/Canada.
AW	Armstrong-Whitworth & Company, Newcastle, Tyne and Wear.
BBC	Brown-Boveri et Cie., Switzerland.
BCK	Bury, Curtis & Kennedy, Liverpool, Merseyside.
BE	Brush Electrical Engineering Company, Loughborough, Leicestershire.
BLW	Baldwin Locomotive Works, Philidelphia, Pennsylvania, USA.
BMR	Brecon Mountain Railway Company, Pant, Merthyr Tydfil, Wales.
BP	Beyer Peacock and Company, Gorton, Manchester.
BRCW	Birmingham Railway Carriage & Wagon Company, Smethwick, Birmingham.
Brod	PRVA Jugoslovenska Tvornica Vagona, Slavonski Brod, Yugoslavia
BTH	British Thomson-Houston Company, Rugby, Warwickshire.
CE	Clayton Equipment Company, Hatton, Derbyshire.
Chr	Fabryka Locomotyw Im F. Dzierzynskiego, Chrzanow, Poland
Corpet	Corpet, Louvet & Compagnie, Seine St. Denis, France
Cravens	Cravens, Darnall, Sheffield, South Yorkshire.
DC	Drewry Car Company, London.
DK	Dick Kerr & Company, Preston, Lancashire.
Djora	Yugoslavia

Dodman	Alfred Dodman & Company, Highgate Works, Kings Lynn, Norfolk.
EE	English Electric Company, Bradford and Preston.
Electro	Electroputere, Craiova, Romania
EW	E.B. Wilson & Company, Railway Foundry, Leeds, West Yorkshire.
FB	Société Franco-Belge de Matériel de Chemins de Fer, Usine de Raismes, France
Florids	Wiener Lokomativfabrik A.G., Floridsdorf, Austria
Frichs	A/S Frichs Maskinfabrik & Kedelsmedie, Arhus, Denmark
FW	Fox, Walker & Company, Atlas Engine Works, Bristol.
GE	George England & Company, Hatcham Ironworks, London.
GRCW	Gloucester Railway Carriage & Wagon Company, Gloucester, Gloucestershire.
Hack	Timothy Hackworth, Soho Works, Shildon, Co. Durham.
Hartmann	Sachsische Maschinenfabrik, Hartmann AG, Chemnitz, Germany
HC	Hudswell-Clarke & Company, Hunslet, Leeds, West Yorkshire.
HE	Hunslet Engine Company, Hunslet, Leeds, West Yorkshire.
Hen	Henschel & Sohn GmbH, Kassel, Germany
HL	R & W Hawthorn, Leslie & Company, Forth Bank Works, Newcastle-Upon-Tyne.
HLT	Hughes Locomotive & Tramway Engine Works, Loughborough, Leicestershire.
Jung	Arn. Jung Lokomotivfabrik GmbH., Jungenthal, Germany
K	Kitson & Company, Airedale Foundry, Hunslet, Leeds, West Yorkshire.
Kitching	A. Kitching, Hope Town Foundry, Darlington, Co. Durham.
Krupp	Friedrich Krupp Maschinenfabriken, Essen, Germany
KS	Kerr Stuart & Company, California Works, Stoke-on-Trent, Staffordshire.
Lima	Lima Locomotive Works Inc., Lima, Ohio, USA.
LO	Lokomo Oy, Tampere, Finland
Loco. Ent.	Locomotion Enterprises (1975), Bowes Railway, Springwell, Gateshead, Tyne and Wear.
Manch	Greater Manchester Museum of Science & Industry, Manchester.
MC	Metropolitan-Cammel Carriage & Wagon Company, Birmingham (Metro-Cammell).
Motala	AB Motala Verkstad, Motala, Sweden
MV	Metropolitan-Vickers, Gorton, Manchester.
N	Neilson & Son, Springburn Locomotive Works, Glasgow, Scotland.
NBL	North British Locomotive Company, Glasgow, Scotland.
Niv	Les Ateliers Metallurgiques Nivelles, Tubize, Belgium
NNM	Noord Nederlandsche Maschienenfabriek BV, Winschoten, Netherlands.
NOHAB	Nydquist & Holm A.B., Trollhattan, Sweden
NR	Neilson Reid & Company, Springburn Works, Glasgow, Scotland.
PR	Park Royal Vehicles, Park Royal, London.
PS	Pressed Steel, Swindon, Wiltshire.
Resco	Resco (Railways), Erith, London.
Resita	Resita, Romania.
RR	Rolls Royce, Sentinel Works, Shrewsbury, Shropshire.
RS	Robert Stephenson & Company, Newcastle, Tyne and Wear.
RSH	Robert Stephenson & Hawthorns, Darlington, County Durham.
S	Sentinel (Shrewsbury), Battlefield, Shrewsbury, Shropshire.
Sara	Sara & Company, Plymouth, Devon.
Schichau	F. Schichau, Maschinen-und Lokomotivfabrik, Elbing, Poland.
Science Museum	Science Museum, South Kensington, London
SM	Siemens, London.
SMH	Simplex Mechanical Handling, Elstow Road, Bedford, Bedfordshire.
SS	Sharp Stewart & Sons, Manchester, England and Glasgow, Scotland.
TK	OY Tampella AB, Tampere, Finland
TKL	Todd, Kitson and Laird, Leeds, West Yorkshire.
VF	Vulcan Foundry, Newton-le-Willows, Lancashire.
VIW	Vulcan Iron Works, Wilkes-Barre, Philadelphia, Pennsylvania, USA.
WB	W.G. Bagnall, Castle Engine Works, Stafford, Staffordshire.
Wkm	D. Wickham & Company, Ware, Hertfordshire.
WMD	Waggon und Maschienenbau G.m.b.H., Donauworth, Germany.
WSR	West Somerset Railway, Minehead, Somerset.
YE	Yorkshire Engine Company, Meadowhall, Sheffield, South Yorkshire.

Keep your book up to date with....

Today's Railways UK

The only magazine to contain official Platform 5 Stock Change information to keep this book right up to date.

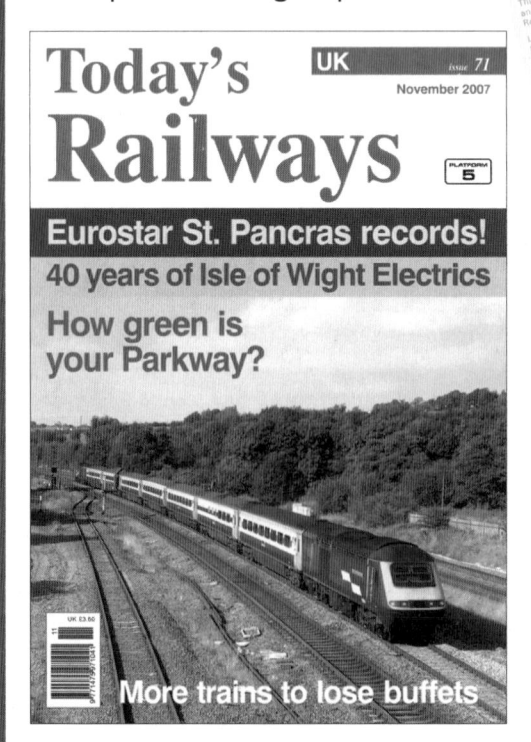

Every issue also includes:

- Latest news about Britain's Railways
- Rolling Stock News
- Light Rail News
- Features + Articles
- Events + Railtours Diary
- Exclusive subscriber discounts

On sale 2nd Monday of EVERY MONTH